Souheila KHALFAOUI

Variography and Sedimentology

Souheila KHALFAOUI

Variography and Sedimentology

Sedimentary and geostatistical characterization of the Lower Devonian reservoirs of the BENOUD Trench (southern Algeria) - A

ScienciaScripts

Imprint
Any brand names and product names mentioned in this book are subject to trademark, brand or patent protection and are trademarks or registered trademarks of their respective holders. The use of brand names, product names, common names, trade names, product descriptions etc. even without a particular marking in this work is in no way to be construed to mean that such names may be regarded as unrestricted in respect of trademark and brand protection legislation and could thus be used by anyone.

Cover image: www.ingimage.com

This book is a translation from the original published under ISBN 978-620-6-71918-2.

Publisher:
Sciencia Scripts
is a trademark of
Dodo Books Indian Ocean Ltd. and OmniScriptum S.R.L publishing group

120 High Road, East Finchley, London, N2 9ED, United Kingdom
Str. Armeneasca 28/1, office 1, Chisinau MD-2012, Republic of Moldova, Europe
Managing Directors: Ieva Konstantinova, Victoria Ursu
info@omniscriptum.com

Printed at: see last page
ISBN: 978-620-8-59161-8

Contents

Glossary

Devonian: Term meaning the second Paleozoic era in the geological time .
Gedinnian: This is the old name for the first Devonian stage, the Lochkovian.
Reservoir rock : Rock where hydrocarbons accumulate (porous and permeable rock).
Parent rock: Sedimentary rock (clay or marl) rich in organic matter. The rock in which hydrocarbons are formed.
Siégenien: This is the old name for the second Devonian stage, the Praguian.
Top: term meaning the top of the sedimentary series.
Permit: Term meaning the exclusive right to explore for oil.
Trap: A structure affecting the reservoir rock and cap rock of an oil system, allowing hydrocarbons to accumulate.

Terms used in petrophysics :

1. K: Permeability.
2. 0: porosity.
3. X: Arithmetic mean.
4. 2: the variance.
5. Standard deviation.

General introduction

The Saharan platform is part of the North African craton. It is bounded to the north by the chains of the Atlas domain, from which it is separated by the southern Atlas flexure, and to the south by the Eglabs and Hoggar Massifs, located to the south-west and south-east respectively. The Saharan platform is subdivided into three sub-provinces: the Western Province, the Eastern Province and the Triassic Province (Figure 1).

The Benoud furrow is part of the western province of the Saharan platform. A great deal of work has been carried out in this region over the last two decades. This work has recently led to a number of promising discoveries that are motivating prospecting in this basin, particularly in the Lower Devonian and Lias reservoirs.

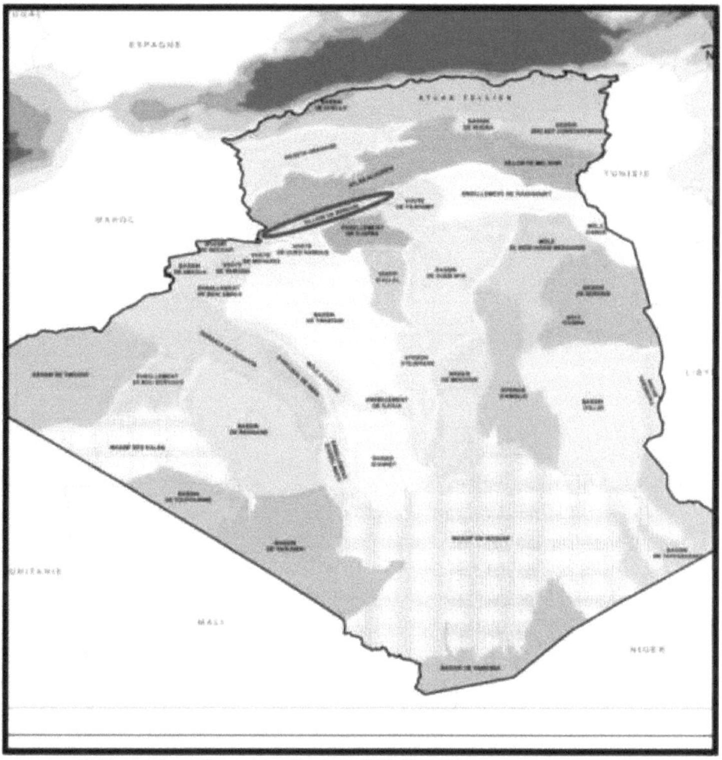

Situation map of Algeria's sedimentary basins, (ALNAFT)

Chapter I

General

+ Problem and objective of the study

This work is part of the collaboration between U.S.T.H. B and Sonatrach (exploration division). It consists of the study and evaluation of the Lower Devonian reservoir in the Benoud Trench. Eight wells have been drilled in our study area: MAMI-1, MAMI-2, MAMS-1, WMAMI-1, AET-1, OAZ-1, GEJ-1 and MAME-1, two of which are new discoveries. The main objective is the Lower Devonian formation, which can be subdivided into three sub-stages: Gedininen, Siégenien and Emsien. They are made up of alternating sandstones and clays. To enable us to carry out a sedimentological approach concerning the Gédininen, Siégenien and Emsien reservoirs, it is necessary to base ourselves on logging data. This approach involves the use of sequence stratigraphy with the interpretation of correlation profiles using "PETREL" software, so it consists first of all in identifying the electrofacies that characterise this interval, in order to see the extension of the sandstone levels, essentially of the Siégenien-Gédinnien type, in the study area.

In order to achieve the desired objective, four chapters have been drawn up:

- The first chapter: Generalities.
- The second chapter : Analysis of the sedimentary fillings of the Lower Devonian.
- The third chapter : Petrophysical analysis and statistical study.
- The fourth chapter : Oil potential.

I.1 History of work in the region

The first research and exploration work in the Benoud Sillon began in the 60s and 70s with the drilling of the TAJ-1 and KEB-1 wells, followed by the drilling of the KHL-1 and EKL-1 wells, located on the Atlas rim.

Drilling activity was relaunched in 2010. The MAMI-1 well (drilled in 2010 in the eastern part of the sillon) was found to contain hydrocarbons following a fracturing operation on a level in the Gedinnian (7.7m3/h of oil and 1934 m3/h of gas). Following this discovery, several wells were drilled on the structure and in the surrounding area, one of which (OAZ-1, in 2018) was found to contain hydrocarbons (oil and condensate, no flow) in a Segennian level. Further west, the TAJ-1 well was drilled again in 2019 and a discovery was made in the carbonate Lias (1.04 m3/h of oil after lifting with nitrogen). In 2020, drilling of the GEJ-1 well resulted in an oil discovery in the Gedinnian with a flow rate of 7.2 m3/h and gas of 3300 m3/h. It should be noted that there are no wells in the central part of the trench. The following table (Table I.1) summarises the history of wells located in the
el Ouabed perimeter and surrounding area:

Well	Year	Depth Final (m)	Floor ACHIEVED	RESULTS
TAJ-1	1965	3802	Cambrian	Gas/oil indices in the Lias Gas indices in the Lower Dev. and Gothlandian
KEB-1	1971	2571.5	Infra Lias	Jurassic acquires
KHL-1	1992	5192	Lower Devonian	Dry
EKL-1	2008	4713	Devonian	Dogger-Lias aquifers
MAMI- 1*	2010	4160	Ordovician	7.7m3 of Oil and 1934 m3 of Gas in Gedinnian after frac

MAME-1	2012	3650	Ordovician	Oil Dev/Ordo (unsuccessful frac)
AET-1	2012	3850	Ordovician	Oil Dev/Ordo (not tested)
MAMI-2	2013	3700	Ordovician	Oil Dev/Ordo (not tested)
MAMS-1	2016	3570	Ordovician	Dry
OAZ-1	2018	4800	Ordovician	Oil and gas in the Siegenian
TAJ-1 Bis	2019	2450	Frasnien	1.04m3/h of oil in the Lias
GEJ-1	2020	4270	Ordovician	7.2m3/h of oil and 3300m3/h of gas in the Gedinnian

Table I. 1- Summary of wells located in the El Ouabed II Perimeter and its vicinity. (Document SH/EXP).

1.2 Geographical location

The Benoud sillón is located in the north-western part of the Saharan platform, south-west of Laghouat. It is a narrow depression oriented WSW-ENE. The El Ouabed perimeter, represented by blocks 103 and 313, covers an area of 15,377.17 km2. It is bounded to the east by the 2°30'E meridian near the town of Aflou and to the west by the 0°30'E meridian passing to the west of el Biod Sidi Chikh. Its northern boundary passes through the town of Tadjrouna, while its southern boundary coincides with parallel 32°30'N passing through the town of Ghardaï'a. (Figure I. 1).

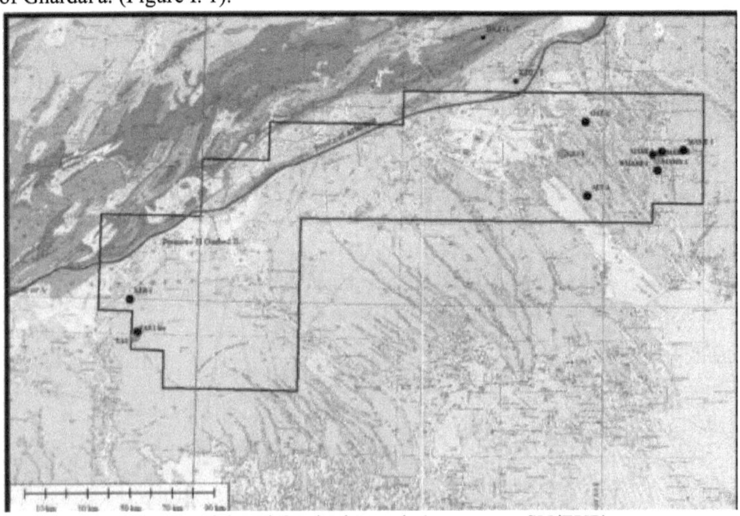

Figure I. 1-Geological map of the El Ouabed Permit (Document SH/EXP)

1.3 Geological setting of the study area (BENOUD basin)

Our study area belongs to the eastern part of the Benoud furrow, which is bounded by: o Western Saharan Atlas to the north.

- The TILRHEMT dome to the south-east.
- The OUED NAMOUS vault to the south-west.

o The DJOFRA silt field to the south and south-east.

The region represents a junction between the Atlasian domain and the Saharan platform.

It forms the transition between a deformed area in the north during the Palaeozoic and Mesozoic-Tertiary periods (Saharan Atlas) and a stable area in the south since the Hercynian (Saharan platform). It is almost entirely covered by Quaternary deposits (Hamada). The only Mesozoic and Tertiary outcrops are located on steep slopes to the north (Figure I.2).

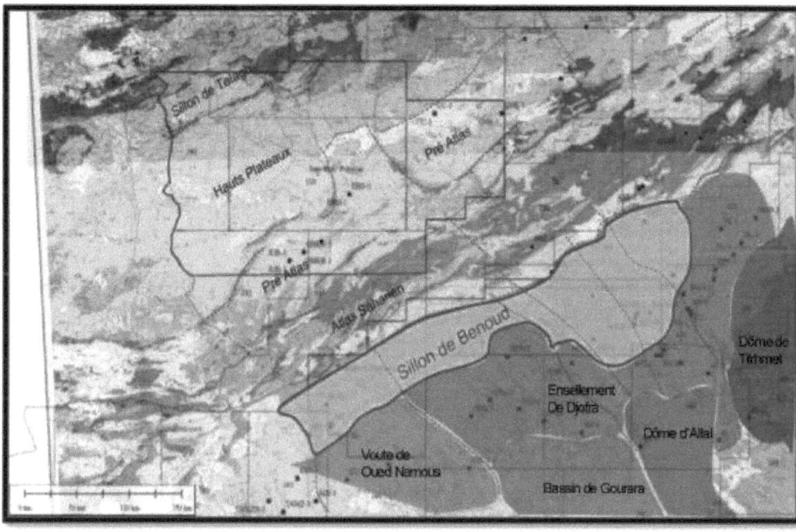

Figure I. 2-Regional structural position of the Benoud Trench. (Document SH/EXP, Sept 2021)

I.3.1 Stratigraphic aspect

The stratigraphic column of the Benoud Trench shows a number of discontinuities, with series ranging from the Cambrian to the Miopliocene, which are essentially detrital.

A. Paleozoic

The Paleozoic sedimentary cover of the Sillón de Benoud is made up of the following series:

Cambrian (Ep>270m)

Cambrian deposits lie unconformably on a crystalline basement and are represented by continental to coastal detrital series.

Cambro-Ordovician

The Bordj Neili and El Megta sandstones are fine quartzite sandstones intercalated with black indurated clays.

Ordovician (330m)

The Ordovician is characterised by detritic sedimentation comparable to that found throughout the northern part of the Saharan platform. It is distinguished by a period of glaciation and alternating transgressions and regressions marked by marine and continental sandstone clays. The Ordovician is characterised by alternating clays and sandstones subdivided into several formations, from bottom to top:

o El Gassi clay ;

o El Anguer sandstone: fine to medium sandstone with clay interlayers;

o Azzel clays: black clays with compact micaceous and pyritic black clays; o Oued Saret clays: grey to black, silty to sandy clays.

o Micro conglomeratic clays: micaceous black clays, silt and gravel with large grains of aeolian quartz;

o The Mkratta slab: quartzite sandstone level.

The Silurian (370m)

The melting of the ice caps caused a marine transgression that affected the entire Saharan platform.

The Silurian is represented by the Oued Ali formation, made up of black organic and radioactive clays with numerous limestone intercalations and sandstone layers.

The Devonian (410m)

In the Lower Devonian to Gedinnian, the sea withdrew from most of the region, giving rise to littoral detritic sedimentation, albeit with some marine influences in the south-west. The Dkhissa formation, which became widespread in the Gedinnian and Segennian, was subdivided into the Lower Clay and Sand Dkhissa and the Upper Sand Dkhissa. At Emsien a new transgression affects the region, the sedimentation becomes clayey with the intercalation of sitestone and quartzite (Tiferguenit formation). At Middle Devonian to Geviitian a warm sea invades the deeper western and south-western regions, where the sedimentation deposited will be represented by carbonates (Sebbat formation). The Frasnian is formed of alternating dolomites and dark clays (Meden-Yahya formation).

The Carboniferous (670m)

The regressive cycle of the Devonian is followed by a transgression in the Carboniferous. The sedimentation is marine (shore face and shelf) with carbonate and detritic deposits.

The following subdivisions can be distinguished during this period:

o The Viséen: represented by marls and clays at the base and carbonates and clays at the top.

o The Namurian: The lower part of the Namurian is characterised by the predominance of black clays, while the upper part is essentially made up of oolitic limestone, sometimes dolomitized.

o Westphalian: alternating fine to medium sandstones, clays and dolomites with limestone layers.

B. The Mesozoic

The Mesozoic of the Benoud sillón lies unconformably on the Palaeozoic. The thickness of the Mesozoic varies between 2000 and 3000 metres, the greatest thickness being found near the southern Atlasian fault.

The Trias (70m)

The whole region is tilted towards the NE, with the Mole Allal still playing a role in the high zone will be the main source of Triassic detrital deposits, although the Hassi R'mel relief will contribute less.

The Trias is formed at the base and summit by red clays interspersed with around forty metres of highly altered green magmatic rocks (dolerites) at the summits.

The Jurassic (900m)

Jurassic includes :

o The Lias: This forms alternating clays and anhydrite at the base, with limestone and clays at the top.

o The Dogger: Marked by a clayey facies at the base which gradually becomes detritic at the top.

o The Malm: essentially composed of alternating sandy clay and sandy clay with dolomitic interlayers.

The Triassic and Jurassic correspond to the Zarzaï'taine formation, which is the sedimentary equivalent.

The Cretaceous (1150m)

The Cretaceous is subdivided into two groups:

o Lower Cretaceous: the Neocomian-Cenomanian period is marked by a clayey-gritty facies sometimes interspersed with clays or marls.

o Upper Cretaceous: in the Turonian-Senonian period, sedimentation becomes carbonate.

C. The Cenozoic

The Cenozoic rests unconformably on the Cretaceous levels; the most representative term of the Cenozoic is the Mio-Pliocene (Hamada formation), with sands and sandstones and intercalations of clay and limestone. In the El Kohol region, on the other hand, the Cenozoic is represented by the Eocene, with sand associated with limestone and clay layers. (Figure I.3).

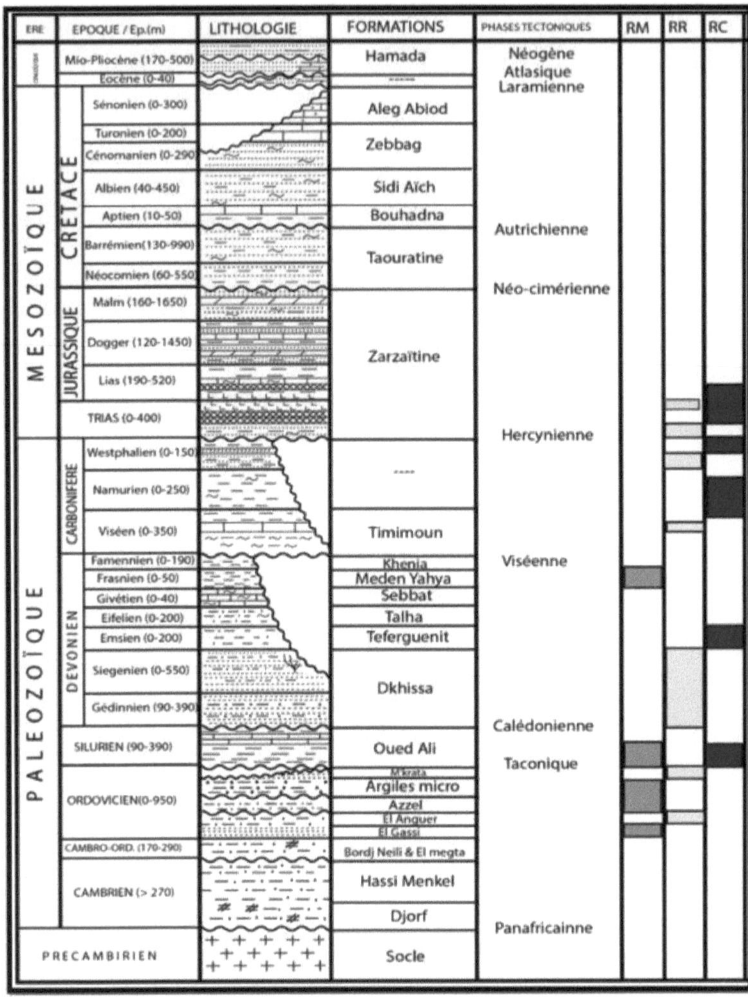

Figure I. 3- Typical stratigraphic section of the Benoud trench. (Document SH).

I.3.2 Structural aspect

The current structural configuration is the result of the influence of several tectonic phases linked to Hercynian and Alpine events. The El Ouabed study area is made up of two structural entities that have evolved differently over geological time: the Saharan Atlas to the north and the Benoud Sillon, part of the Saharan platform, to the south. These two structural units are separated by a major fault that crosses the whole of North Africa from west to east. According to the authors, this fault is known as the South Atlasic Overthrust Front (SAF).

I.3.3 Geodynamic evolution

Pan-African event

At around 620 Ma, several ancient cratons came together to form the mega continent Gondwana. In North Africa, these were the West African craton and the East African craton.

East-west compression resulted in the formation of large submeridian rifts underlined by mylonites, combined with dextral NE-SW and senestial NW-SE faults (Boudjema, 1987.Haddoum, 1992).

Taconic Event

This event began in the Areenigian and led to the uplift of the Hassi R'mel and Oued Namous domes, causing the series to thin towards these high areas. The Ordovician is characterised by local unconformities, large-scale folds and, above all, a southward regional tilt, resulting in severe erosion on the northern flank (Oued Namous and Hassi R'mel) compared with the southern flank (Oued Mya) where the Ordovician series is complete. The Taconic unconformity itself is described at the base of micro conglomeratic clays of Asghillian age.

New Caledonian event

The Caledonian event is linked to the closure of the Iapetus Ocean, accentuating the compressive movements that lifted the Saharan platform during Taconic event.

The unconformity brought together deep-sea marine clays (Oued Ali series) of Silurian age and Devonian sandstones (Dkhissa sandstone).

Hercynian event

From the Devonian to the end of the Carboniferous, this event led to the collision of the mega-continents Gondwana and Laurasia to form the mega-continent Pengea (Andrako, 1993).

In the Viséen, a N040° compressive phase (Boudjema, 1987) produced NW-SE structures and major uplifts in the region, followed a period of erosion.

In the Namurian, the effects of this phase, which are only visible in the south-western part of the Benoud Trench, are manifested by an unconformity that brings the Namurian over terrain ranging from Upper Devonian to Lower Viséan.

Ante-Westphalian, the Westphalian lies unconformably on the Devonian to the north and north-east of the Benoud trench. This unconformity extends from the Béchar region, where the Moscovian overlies the Lower Bashkirian or Serpukhovian (KaziTani and Durée, 1993), to the Bordj Nili region, where it overlies the Silurian.

In the Post-Stephanian a compressive phase trending N120 (Boudjema, 1987) produced NE-SW structures and major uplifts (Anadarko, 1993) followed by a major cycle of deep erosion of the Palaeozoic, locally up to Silurian (Mazar palaeo-uplift), resulting in a regional unconformity separating the Palaeozoic from the Mesozoic.

The Alpine event

The history of the Benoud Trench during the Alpine event (Meso-Cenozoic) is directly related to the opening of the Atlantic that began in the Triassic period.

The Sillon de Benoud was formed in two phases:

Triassic -Lower Jurassic

The Benoud region at this time was a high zone, the result of a system of normal faults on the southern shore of the Atlas Basin, forming a half-graben.

Middle Jurassic - Present

The Benoud Trench began to subside in the Middle Jurassic, with an irregular rate of subsidence. Subsidence slowed down in the Cretaceous and accelerated in the Tertiary. The subsidence mechanism of the Benoud Trench can be linked to the development of a passive sedimentary margin beneath the area currently occupied by the Atlas .

The distensive phase of the Trias-Lias

During the Triassic-Liassic period, the Benoud Trench was located on the southern margin of the Atlas Trench. The distensive regime caused replay of ancient NE-SW faults into normal faults, resulting in facies changes and thickness variations along these faults and the deposition of volcanic rocks that are found today in the Benoud Trench and neighbouring regions (TAJ- UC-101) (Anadarko, 1993).

The Austrian phase

Reverse faulting of the submeridian faults in the Benoud region led to the formation of north-south trending anticlinal vaults. These were subsequently affected by intense erosion, although only in the Barremian. This phase is recognised seismically by an unconformity (the Austrian unconformity) separating the terminal Aptian from the Barremian, and by the essentially reverse faults truncated by this unconformity.

The Laramian phase

The differentiation of the Benoud Trench probably began with this phase, in the Upper Cretaceous. During this event, the Saharan Atlas emerges and at its foot, on the Saharan side, molassic gutters are established, including the Benoud Trench, receiving the products of the dismantling of the Saharan Atlas (KaziTani, 1986; KaziTani and Duée, 1993). This phase affects the Cenomano-Turonian series which outcrop in the El Kohol region and is expressed by folds oriented N330 as well as tension cracks in echelons indicating N240 shortening (Habani and Medaouri, 1994).

The major Atlas phase

This Eocene phase, which is expressed by reverse faults and mainly NE-SW folds indicating a NW-SE squeezing direction, is responsible for the uplift of the Atlasic domain. The uplift was followed by erosion, which is now expressed at the subsurface by a regional unconformity (Atlasic unconformity) separating the Mesozoic (Cretaceous) from the Cenozoic (Mid-Pliocene).

The Neogene phase

This phase is notably responsible for the formation of South to South-East dipping thrusts at the front of the Atlas Range, at the boundary with the Benoud Trench, surmounted by narrow NE-SW folds. Outcrops in the El Kohol region show the formation of dextral NW-SE and senestial NE-SW thrusts compatible with North-South compression, which sometimes offset pre-existing structures.

Chapter II

Analysis of the sedimentary fillings of the Devonian.

11.1 Introduction

In order carry out a sedimentological study of the Gedinnian-Siegenian and Emsian reservoirs, it is necessary to rely on logging data. This approach involves the use of sequential analysis for the reference well (MAMI-1) and sequential stratigraphy for all the wells with the interpretation of correlation profiles, so it consists first of all in identifying the elevrofacies, we used the logging results produced by Sonatrach in the study area, on the wells; MAMI-1, AET-1, GEJ-1, OAZ-1, MAME-1(Figure II .1), and also a description of the core samples from the five wells, with MAMI-1 taken as the reference well. In order to have a good representation of the spatial evolution of the depth of these reservoirs in the study area, isobath maps were drawn up. To process all the digital data, we used Pétrel software, version 2017.

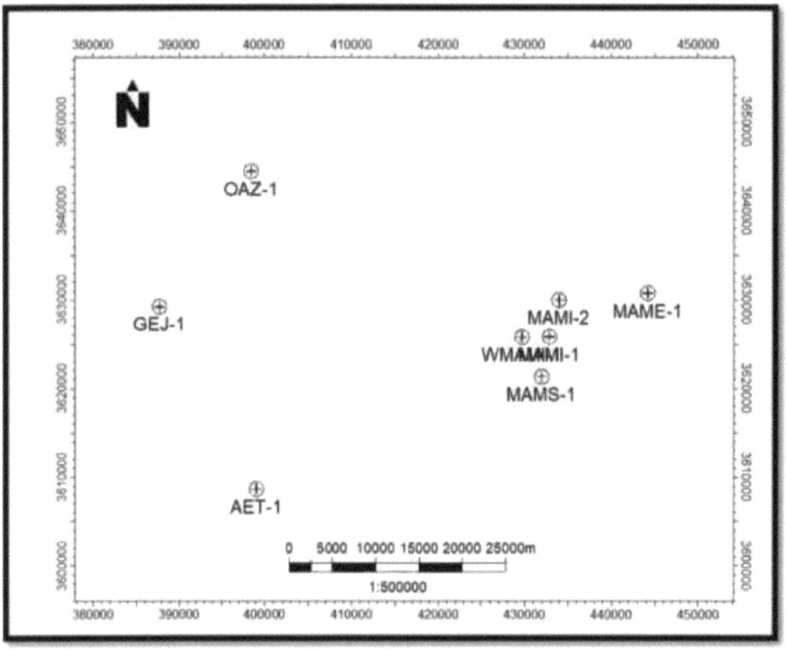

Figure II. 1-Map showing the location of the wells studied

Sequences and sedimentary processes are identified mainly by applying the principles of sequence stratigraphy to the log record, in particular those of Gamma-Ray. This identification is literally based on highlighting the different points of inflection of the Gamma-Ray curve and the gradual or abrupt variations in this curve corresponding to the passages from one sequence to another and from one procession to another.

The relationship between changes in grain size, argillosity and the behaviour of the GR and DT curves has made it possible to classify the log curves according to their shape (Figure II.2).

+ The bell shape: related to fluvial channel deposits or transgressive deposits, its basal contact is abrupt.

+ The funnel shape: characterises a middle of bars with a gradual basal level.

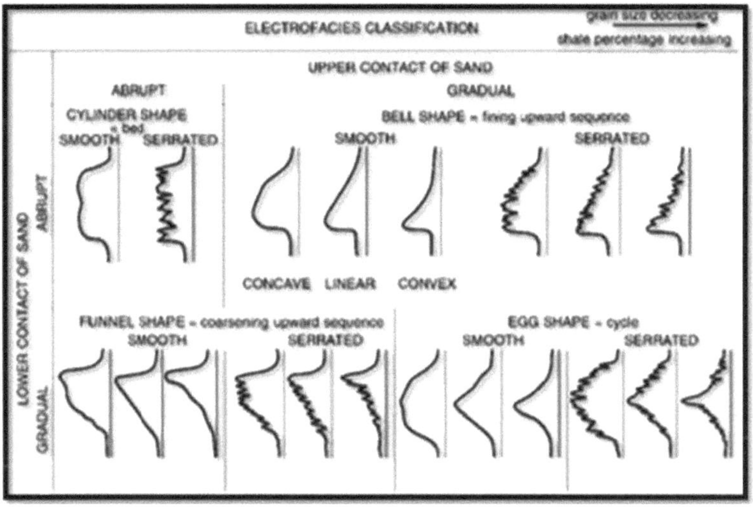

Figure II. 2-Classification of electrosequences (after SERRA and SULPIICE, 1975)

11.2 Notion of sequence stratigraphy

Sequence stratigraphy is a discipline that brings together a number of methods with aim of gaining access to the geometric and chronological relationships of stratigraphic units within sedimentary assemblages. Its aim is to define sedimentary units bounded by time-valued surfaces that correspond to particular periods of sea-level variation, based on the analysis of seismic sections, log records, outcrops and data from core descriptions.

Sequence stratigraphy is an excellent tool for identifying and quantifying the ways in which sedimentary basins are filled. Its concepts were developed by a group of geophysicists and petroleum geologists on the basis of seismic stratigraphy.

11.3 Concept of sedimentary processes

A sedimentary procession (tract system) is defined by a sequence of contemporary depotional systems (Brown Jr and Fischer, 1977). Each procession is associated with a particular segment of the eustatic curve. Three main processions have been identified: (Figure II.3)

a. The Lowstand Systems Tract (LST)

It is limited at the top by the transgression surface. This procession is deposited during the rapid fall in sea level. In detail, the low-level procession is subdivided into three :

- Bassin floor fan (BFF) or underwater cone.
- Slop fan (SF) or underwater slope cone.
- Low prograding wedge or complex prograding wedge.

b. The Transgressive Systems Tract (TST)

It is deposited during the eustatic cover and this interval either overlies the low-level deposits or directly overlies the high-level deposits. It is bounded at the base by the first marine transgression surface (ST) and at the top by the maximum flooding surface (MFS). It is marked by retrograde parasequence sets. In logging, it is represented by a stack of geological layers giving the shape a bell which becomes smaller towards the roof.

c. The Highstand Systems Tract (HST)

This procession is formed during the eustatic ascent and is located in the upper part of the sequence. It begins with aggrading parasequences and ends with a prograding parasequence, developing towards the end of the eustatic ascent during eustatic stability and during the beginning of the fall in sea level.

11.4 Discontinuity surfaces

On a basin scale, genetic sequences can provide certain markers with chronological values (surfaces of discontinuities) that can be used as references for correlations. These surfaces incorporate a certain time interval, of variable importance depending on their nature and position in the basin:

a. Sequence limits (LS or SB)

There are two main types of sequence limits (LS or SB):

□ The LSI or SB1 discontinuity forms when the fall in sea level is greater than that of subsidence. It limits a depositional sequence (made up of LST, TST and HST) towards the base.

□ The LS2 or SB2 discontinuity forms when the eustatic fall is equal to or slightly less than that of the subsidence. It is limited towards the base by a depositional sequence composed of a Shlef Margin System Tract (SMST), a transgressive system tract (TST) and a High Stand System Tract (HST).

b. First flooding surface (ST transgression surfaces)

This is a flooding surface, corresponding to the first flooding surface through which the deposition of the retrograde transgressive procession below the low sea level procession or platform edge begins.

c. Maximum Flooding Surface (MFS)

This surface marks the boundary between the transgressive procession and the high sea level procession, a condensation surface (condense section).

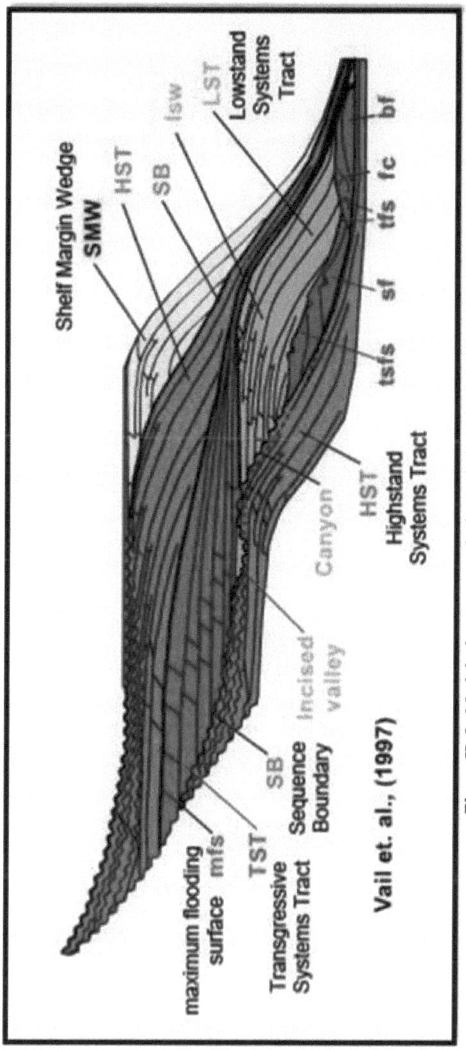

Figure II. 3- Fecole deposition sequence model from vail et al, (1997).

11.5 Lithological description of the Lower Devonian cores

Exceptional conditions this year did not allow us to access the cores, but we were nevertheless able to benefit from core descriptions made by Sonatrach. (SONATRACH, April 2012) We chose five wells for the core descriptions: MAMI-1, AET-1, GEJ-1, OAZ-1 and MAME-1.

MAMI-1 well

Three cores are located in the MAMI-1 well:

Core N °05 lies between 3121.5m and 3105, and is 16.5m thick. This interval is presented by grey-black, indurated, silty, micaceous clays (Figure II.4)

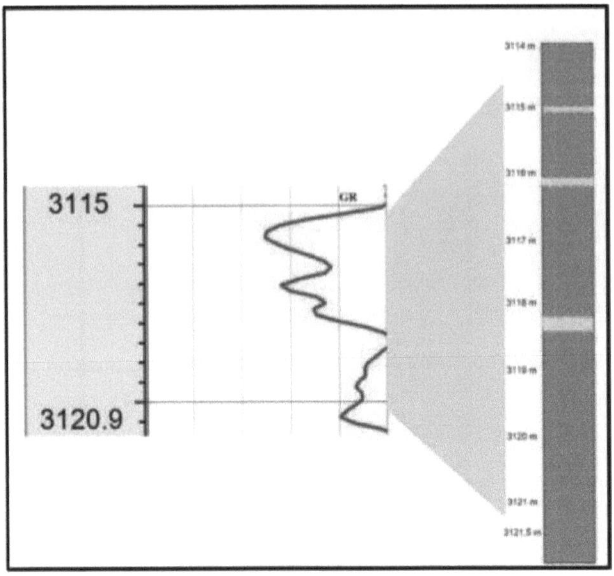

Figure II. 4- Illustration of the clays in core N °05(interval of 3121.5m to 3105m).

Core N °04 from 2822.5 to 2840.5m This interval is a grey-white to white sandstone, fine to medium, sometimes very fine, siliceous to silico-carbonate, moderately hard to hard, micaceous, with low to medium porosity, showing spots of pale yellow direct fluorescence at the top (2822.5m), with veinlets and very fine layers of grey siltstone, clayey, hard, micaceous.

Core N °03 from 2725 to 2731m This interval consists of a siltstone complex composed of: dark grey siltstone, quartz clay, indurated, very micaceous, with centimetric to decimetric passage, nodules and veinlets of grey-white sandstone, very fine, compact, hard, siliceous, subangular, micaceous, pyritic in places, rare oblique fissures with siliceous filling.

Well AET-1

Three cores are located in the AET-1 well**:**

Core N °03 lies between 3112m - 3130m and is made up of :

White to beige, locally grey, greenish, fine siliceous to silico-carbonate sandstone, consolidated with centimetric layers of dark grey, indurated, silty, micaceous clay. Presence of friction mirrors and fossil imprints.

Core N °02 lies between 3036m and 3054m and is made up of white to beige, fine to rarely medium, sub-rounded, silico-clayey, moderately consolidated sandstones.

Core No. 01 lies between 2740m and 2758m and consists of a grey to dark grey, indurated, silty to micaceous, pyritic siltstone, sometimes changing to very hard dark grey siltstones with millimetre to centimetre passages of light grey to white, fine to very fine, clayey to siliciclastic, hard compact sandstone.

Well GEJ-1

Core No. 01 lies between 3536 m and 3554 m and is characterised by dark grey, fine to very fine, clayey, crumbly to moderately hard sandstone.

OAZ-1 well

Three cores were found in the OAZ-1 well:

Core N °04 is located at the level of sequence A between 4153 m and 4135m and consists from bottom to top of :

Grey-white to dark-grey sandstone, fine in places, medium silico-quartzite to silico-clay, medium hard with a few banks of dark grey to grey-black, silty, micaceous clay.

- Dark grey to grey-brown sandstone, very fine, silico-clayey, moderately hard.

Core N °03 is found in sequence C between 4054 m and 4036 m and consists of a complex of very fine, dark grey, silico-quartzitic, hard, compact, white-grey sandstone and dark grey, highly silty, micaceous, fossiliferous, indurated clay.

Core No. 02 is located in sequence D between 3701 m and 3683 m, consisting of the following from bottom to top:

□ Grey to dark grey, silty, soft to indurated clay with fine layers of beige-white, very fine, silico-clayey, moderately hard sandstone and grey-white to white, chalky, moderately hard limestone.

□ Grey to dark grey, silty to strongly silty, indurated clay with centimetric lenses of very fine, quartzite, hard and compact grey-white sandstone.

MAME-1 well

Eight cores are located at the MAME-1 well:

Lithological interval	No. of carrots	Lithological description and sedimentary structures
3286m -3304m	N °10	Consists of very fine, brown to dark grey, soft silico-clay sandstone, with thin patches of light grey clay, Fl fluorescence: pale. Visual porosity: low.
From 3212 to 3221m	N °9	Made up of grey-white to white sandstone, very fine to fine, passing to siltstone, subangular to subrounded, siliceous to silico-clayey, moderately consolidated to friable, micaceous, with rare fine passages, grey to grey-black clay, indurated, silty, micaceous (Figure II.5).
From 3212m to 3194m	N°8	Consisting of, from bottom to top : □ Grey to grey-white sandstone, very fine to fine, subangular to subrounded, siliceous to silico-clayey, moderately hard, micaceous with rare fine layers of grey to grey-black clay, indurated, silty, micaceous. □ Grey to grey-black clay, indurated, silty, micaceous, with fine centimetric to filmy layers of very fine to fine grey-white to white sandstone, silico-clayey, hard, micaceous
From 3181m to 3163m	N°7	Consists of grey to grey-black clay, indurated, silty, micaceous, with fine centimetric layers to films of grey-white to white sandstone, very fine to fine silico-clay, hard, micaceous, sedimentation in the form of millimetre and centimetre horizontal laminations.
From 3145m to 3127m	N°6	Characterised by: grey to light grey clay, grey-black, soft to indurated, silty, micaceous, with intercalations of grey-

		white to white sandstone, very fine to fine, silico-clayey, hard, micaceous.
From 3038m to 3020m	N°5	Consisting of, from bottom to top : □ Clay-sandstone complex composed of : White to white, very fine to fine, silico-clayey, hard, micaceous sandstone with centimetric to filmy layers of black, indurated, silty, micaceous clay. □ Grey-white to white sandstone, very fine to fine, silico-clayey, hard, micaceous with centimetric to filmy layers of black clay, indurated, silty, micaceous.
From 2799m to 2789m	N°4	Made up of very fine dark grey silico-clayey hard sandstone with layers of black silty indurated clay and hard grey-black silstone (Figure II.6).
From 2789m to 2771m	N°3	Consisting of, from bottom to top : □ Fine to very fine dark light grey silicic clayey sandstone with centimetric and decimetric levels of indurated black silty clay, micaceous. □ Light grey fine to very fine siliceous sandstone

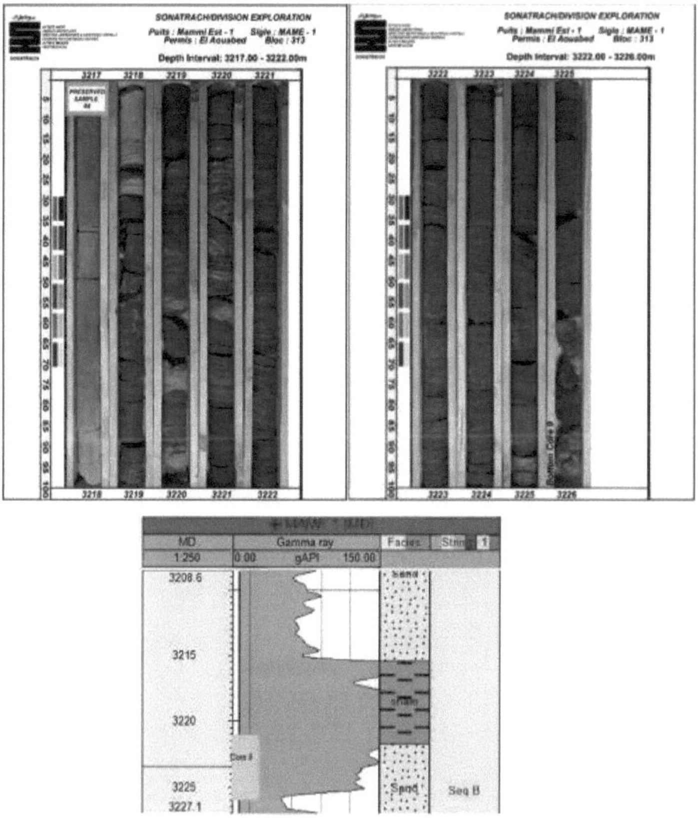

Figure II. 5 - Photos of core 09 (interval 3226m-3217m) from the MAME-1 well.

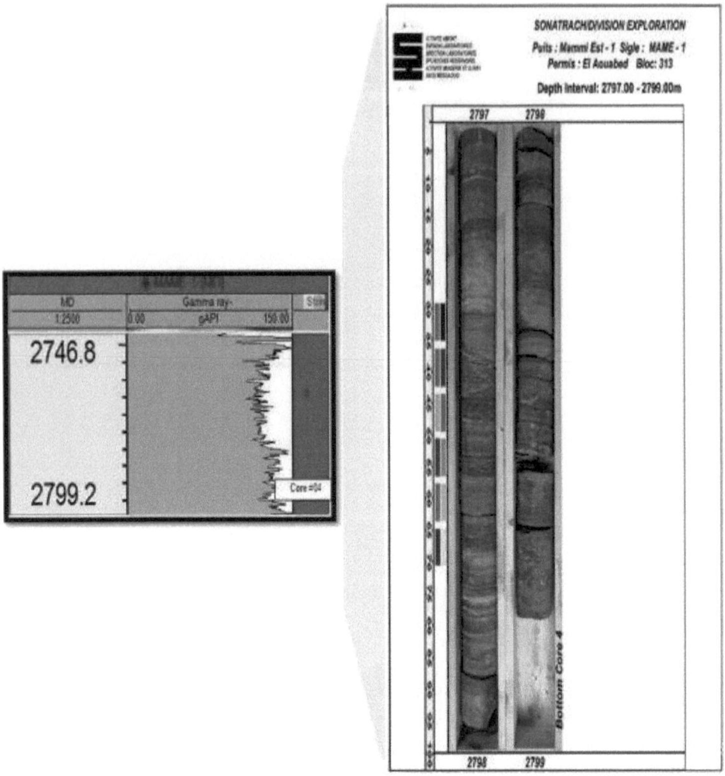

Figure II. 6 .Photos of core 04 (interval 2799m-2797m) from the MAME-1 well.

II.5.1 Core analysis

The description of the cores in holes MAMI-1, MAME-1, OAZ-1, GEJ-1 and AET-1 shows the presence of six essential facies:

1. Facies : Silty clay with very fine sandstone.
2. Facies: Fine to very fine sandstone with fine clay.
3. Facies. Very fine to fine sandstone.
4. Facies. Clayey-gritty with bioturbation.
5. Facies: Obliquely stratified silt-gravel complex.
6. Sandstone, silt and clay interlayers.

In what follows, we provide a detailed description of these facies and their interpretation.

1. Facies Silty clay with very fine sandstone layers

Silty clay with very fine sandstone. Dark grey clay, indurated, silty to siltstone in places, strongly micaceous, with nodules and fine sandstone layers. Sedimentation in the form of millimetre and centimetre horizontal laminations.

This predominantly clay facies is attributed to the offshore environment.

2. Facies: Fine to very fine sandstone with fine clay.

Grey-white to white, very fine to fine, silico-clayey, hard, micaceous sandstone with centimetric to filmy layers of black, indurated, silty, micaceous clay. This facies is interpreted as being deposited in a shallow marine environment under middle shoreface to upper shoreface conditions dominated by storms due to the presence of fine clay interbeds between the sandstone banks.

3. Facies: Very fine to fine sandstone.

White sandstone, very fine to fine, sometimes medium, carbonated, moderately to well consolidated, micaceous, slightly bituminous. Grey-black, indurated, silty, micaceous clay. This facies was probably deposited in the lower shoreface because of the predominance of a very fine grain size in the sandstone beds.

4. Facies: clayey-gritty with bioturbation

White to beige sandstone, locally greenish grey, fine, siliceous to silico-carbonate, consolidated with centimetric layers of dark grey, indurated, silty, micaceous clay. Presence of friction mirrors and fossil footprints. Lower shoreface due to bioturbation.

5. Facies: Obliquely stratified silt-gravel complex

Silt-gravel complex composed of grey to dark grey siltstone, clayey with indurated quartz grains, very micaceous with centimetric passages of nodules and veinlets of grey-white sandstone, very fine, compact, hard, siliceous, sub-angular, micaceous, pyritic in places and rare oblique fissures with siliceous filling. Upper shoreface facies.

6. Sandstone, silt and clay interlayers

White to beige sandstone, locally grey, greenish, fine siliceous to silico-carbonate, consolidated with centimetric layers of dark grey clay, indurated, silty, micaceous. Presence of friction mirrors and fossil imprints.

II.6 Sequencing

We chose five wells for sequential cutting: MAMI-1, MAME-1, OAZ-1, GEJ-1 and AET-1. By correlating with the typical Beicip cross-section (Figure II.7), we have established the sequential sectioning for the five wells. We have detailed the sectioning only for the MAMI-1 well (reference well) and the other wells are based on the same principle.

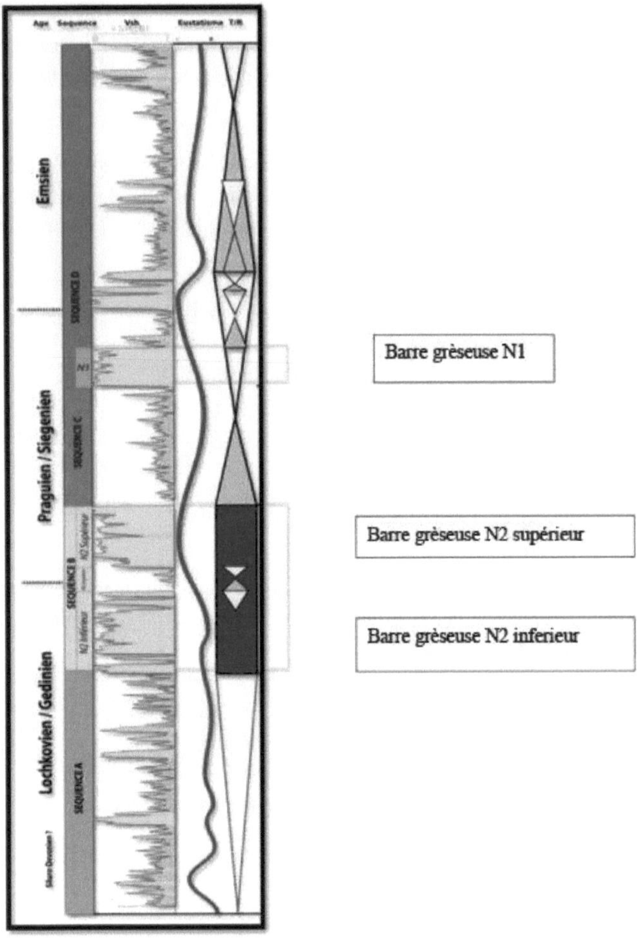

N1 bar grader
Upper N2 gravel bar
Lower N2 grading bar

Figure II. 7- Sequence division of the Lower Devonian of the Sillón de Benoud From Beicip (2013)

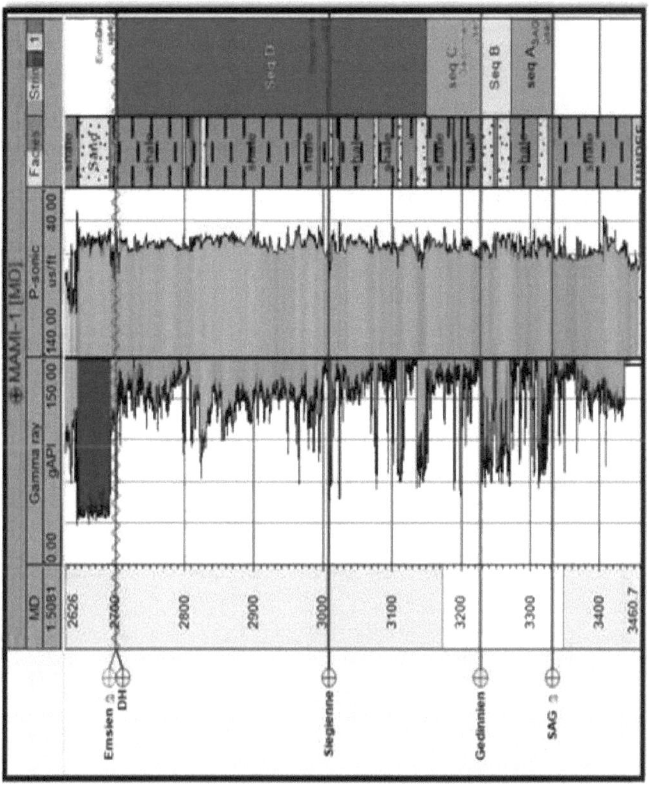

Figure II. 8-Lower Devonian sequences from the MAMI-1 well (based on Fig.

II.7 Sequential analysis of the MAMI-1 well using the Beicip type section

Sequence A of the MAMI-1 well

This sequence (Figure II.9) is delimited at its base by the SAG top (Silurian Clayey Greasy) and is made up of four greasy levels interspersed with clays. It is approximately 60.06m thick.

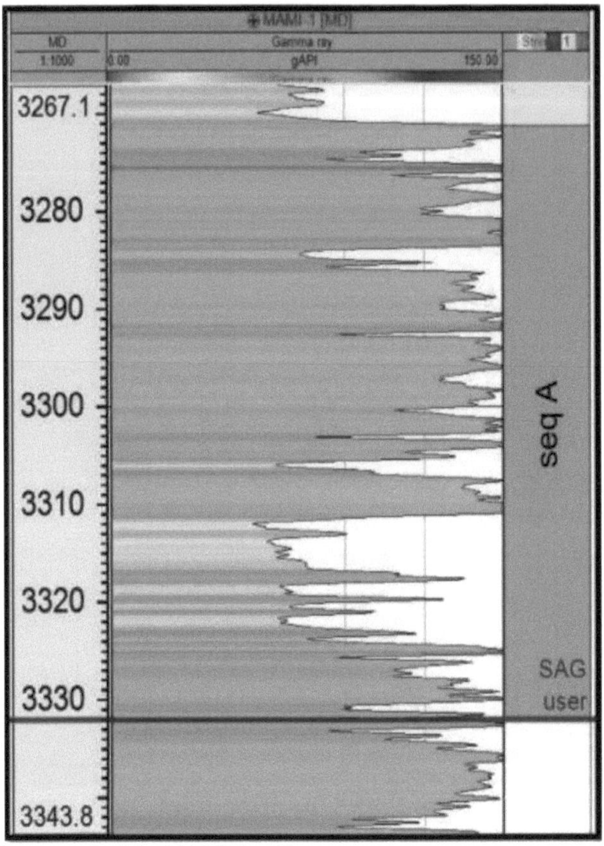

Figure II. 9- Sequence A of the MAMI-1 well

Sequence B from the MAMI-1 well

This sequence (Figure II.10) is delimited at the top by the Gedinnian top and consists of two sandstone bars, the last two of which are separated by a thin clay bank. It is about 44.45m thick.

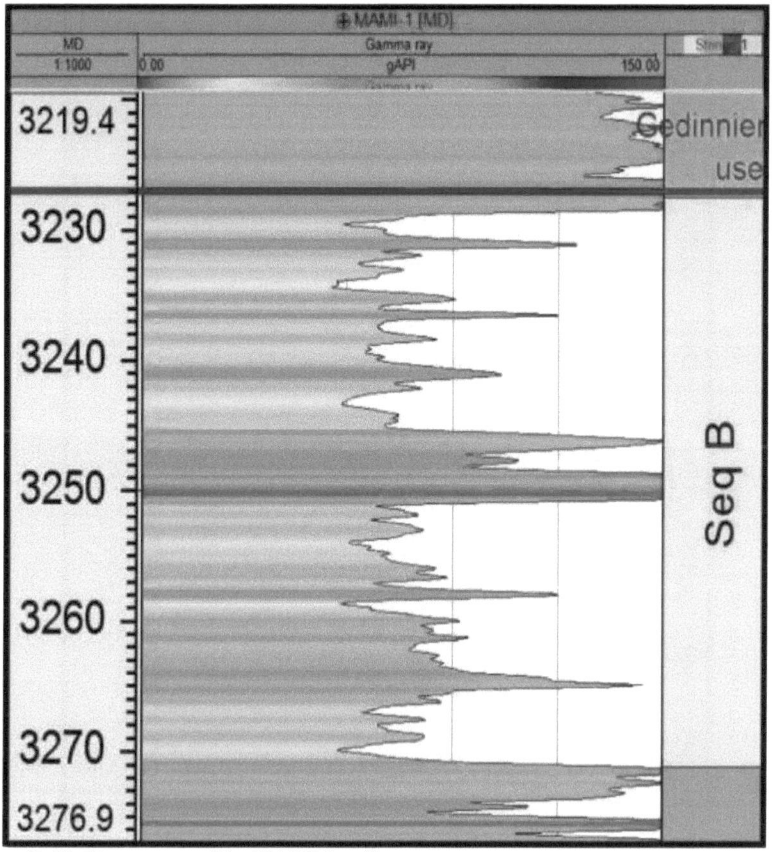

Figure II. 10- B sequence of the MAMI-1 well

Sequence C of the MAMI-1 well

With a thickness of 72.76 m, limited at the base by the Gedinnian top, it is represented by a clay bed, with some sandstone in between. (Figure II.11)

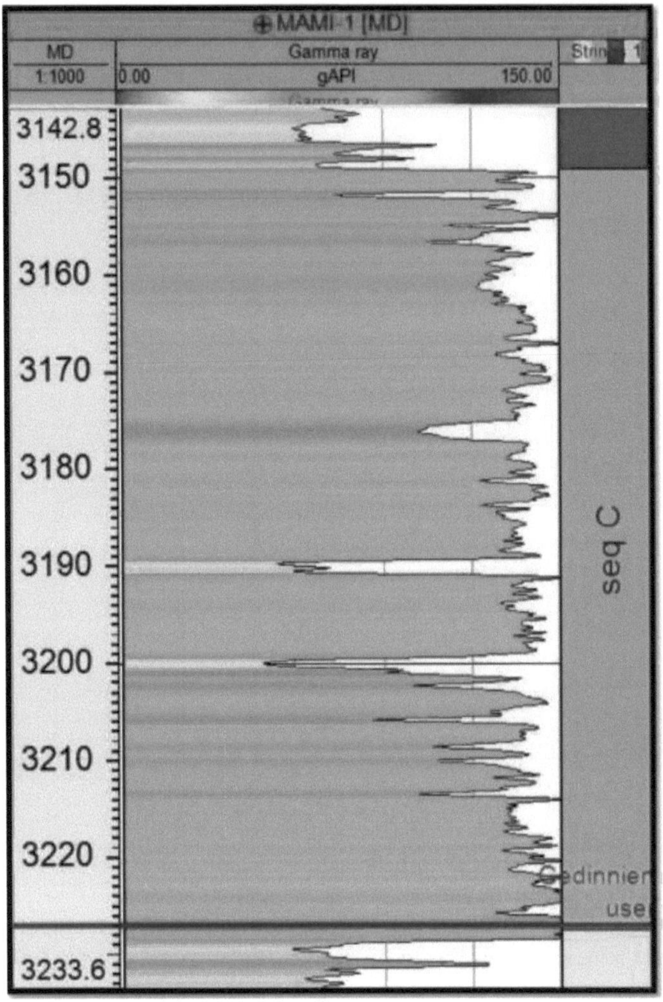

Figure II. 11- C sequence of the MAMI-1 well

Sequence D of the MAMI-1 well

449.53m thick, represented by alternating sandstones and clays, bounded at the base by a sandstone bed and at the top by the Hercynian unconformity. (Figure II.12)

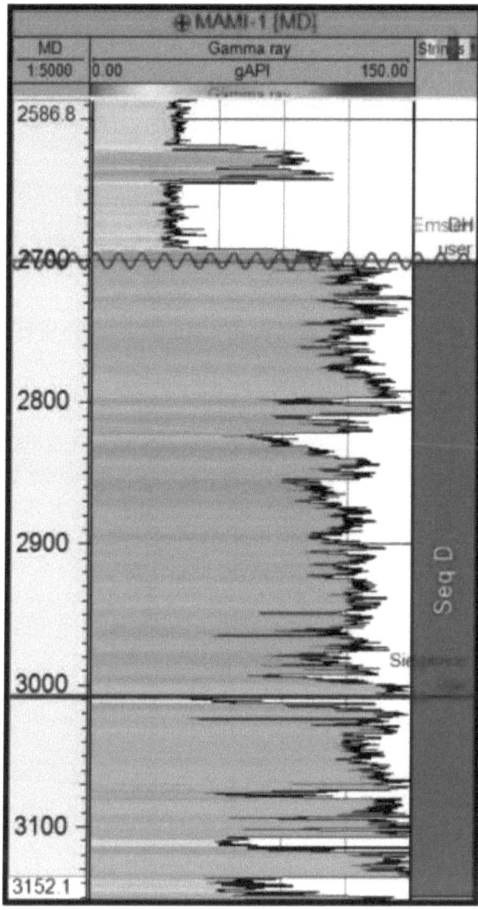

Figure II. 12- D sequence of the MAMI-1 well.

II.8 Sequence stratigraphy

Sequence stratigraphy of the MAMI-1 well

Based on the sequential analysis, four sequences were defined (A, B, C, D) (Based on SONATRACH data)

Sequence A

With a thickness of 60.06m limited at the base by SB (Boundray Sequence), the latter is determined on a Gamma-Ray diagraphic recording by a significant downward inflection of the diagraphic curve. This sequence is characterised by the succession of two (2) sedimentary processes. (Figure II.13)

A low-level prism (LST-1)

This prism shows a reuf-shaped Gamma-Ray curve, which implies the existence of a positive granocroissant evolution, with a prograding architecture due to a drop in sea level (regression), followed by another negative granocroissant evolution and a retrograding architecture caused by a rise in sea level (transgression).

A high-level marine prism (HST-1)

The Gamma Ray diagonal curve has a cylindrical shape with a positive trend.

Sequence B

44.45m thick, limited at the top by a Boundray Sequence, characterised by the succession of three (3) sedimentary processes. Two low marine level processes interspersed with a transgressive process.

Two low-level prisms (LST-2 and LST-3)

The Gamma-Ray recording curve is characterised by a rectangular shape (electrofacies morphology).

A transgressive procession (TST-1)

The GR curve shows a bell-shaped morphology of the electrofacies, implying a positive trend.

Sequence C

Characterised by the presence of eight sedimentary cortèges TST-2, TST-3, HST-2, TST-4 and LST-3, LST-4, LST-5, LST-6. The TST-2, TST-3, HST-2 and TST-4 sedimentary processes are characterised by funnel-shaped electrofacies, with a type of granocroissant sequence with a prograding evolution, represented by clays at the base surmounted by sandstones at the top. It around 72.70 m thick.

Sequence D

449.53m thick, characterised by alternating prisms of low marine level with prisms of high marine level which are more developed at this level with some passages of transgressive processes.

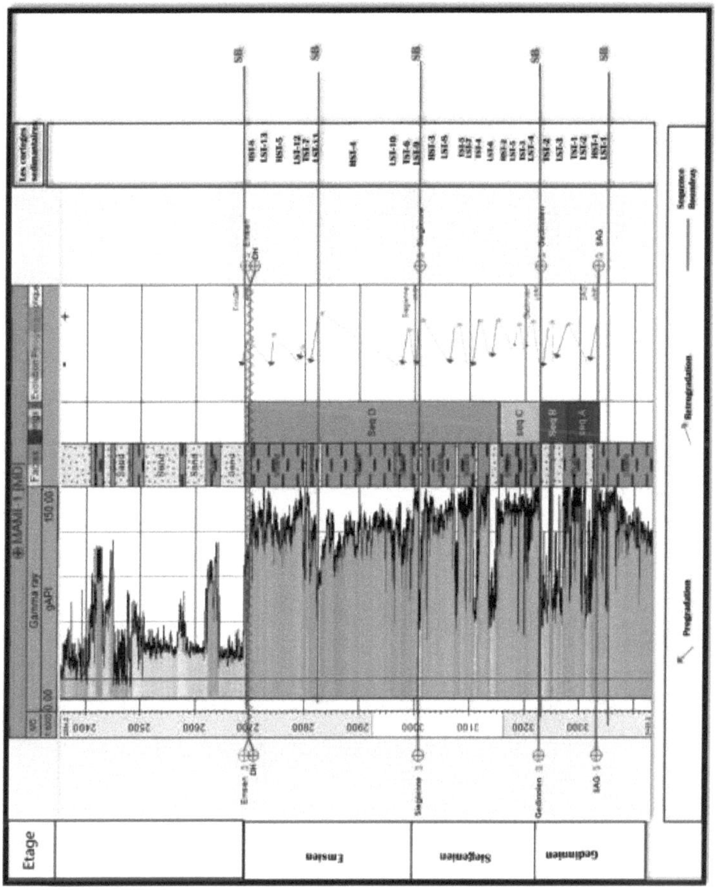

Figure II. 13- Sequential sectioning of the MAMI-1 well

Sequence stratigraphy of the AET-1 well

Four sequences have been defined:

Sequence A

41.81m thick, bounded by two sedimentary prisms, a TST (TST-1) followed by an LST (LST-1) (Figure II.14).

Core 04 lies between 3235.35 and 3240.5m, is 18m thick and consists of black carbonaceous, indurated, silty, micaceous clays with abundant centimetric nodules of siltstone.

Sequence B

49.22m thick, characterised by two sedimentary prisms, a TST (TST-2) followed by an LST (LST-2).

Sequence C

8.91 m thick, characterised by a succession of two sedimentary processes, TST-4 followed by LST-3.

Sequence D

498.65m thick, characterised by a succession of high sea level prisms (HST-1 to HST-6) and low sea level prisms (LST-5 to LST-15), bounded at the top by a Boundray Sequence.

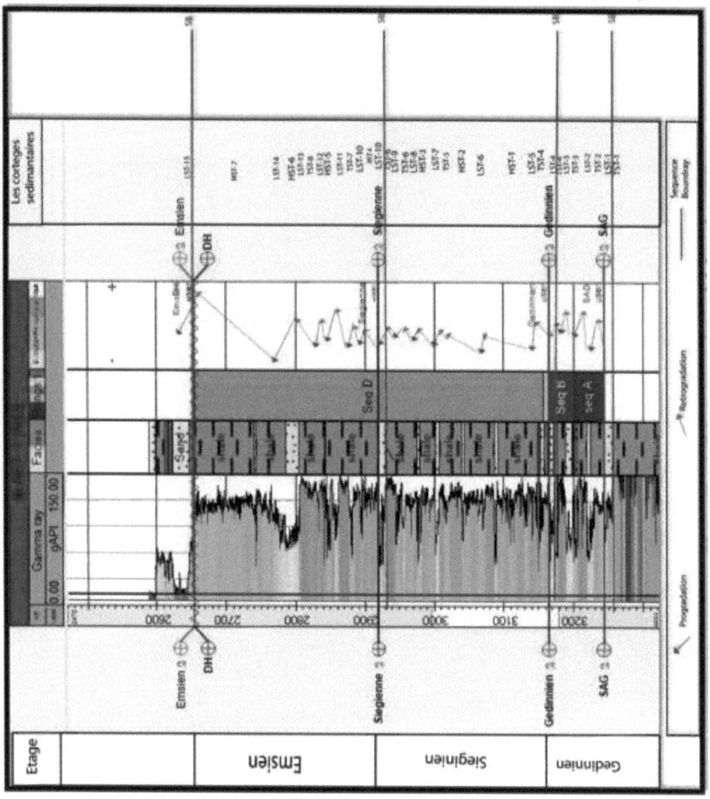

Figure II. 14- Sequential breakdown of the AET-1 well.

Sequence stratigraphy of well GEJ-1

Four sequences have been defined:

Sequence A

62.01m thick, bounded at the base by a SB, characterised by a high marine level prism HST-1 (Figure II.15).

Sequence B

Its thickness is around 14.7m, characterised by a low-level marine prism LST-1 limited at the top by a SB, this is a sandstone bed.

Sequence C

Its thickness is around 37.77m, characterised by a high-level marine prism HST-2.

Sequence D

With a thickness of 354.4 m, it alternates low-level marine prisms with high-level prisms.

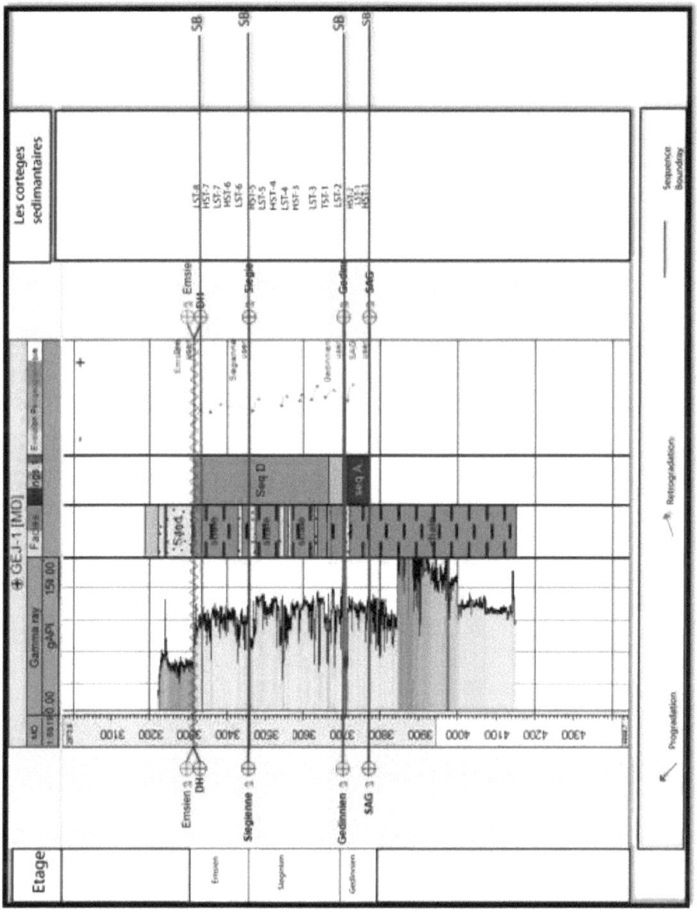

Figure II. 15 - Sequential sectioning of well GEJ-1.

Sequence stratigraphy of the OAZ-1 well

Four sequences have been defined:

Sequence A

85.33 m thick, characterised by the presence of the following sedimentary processes: LST-1, TST-1, LST-2, TST-2, LST-3, HST-1, LST-4, HST-2: LST-1, TST-1, LST-2, TST-2, LST-3, HST-1, LST-4, HST-2 (Figure II.16)

Sequence B

42.33m thick, succession of three sedimentary processes LST-5, TST-3, LST-6.

Sequence C

47.63m thick, characterised by a single HST-3 sedimentary process.

Sequence D

347.9 m thick, characterised by alternating low-level prisms and high-level prisms with some transgressive processions.

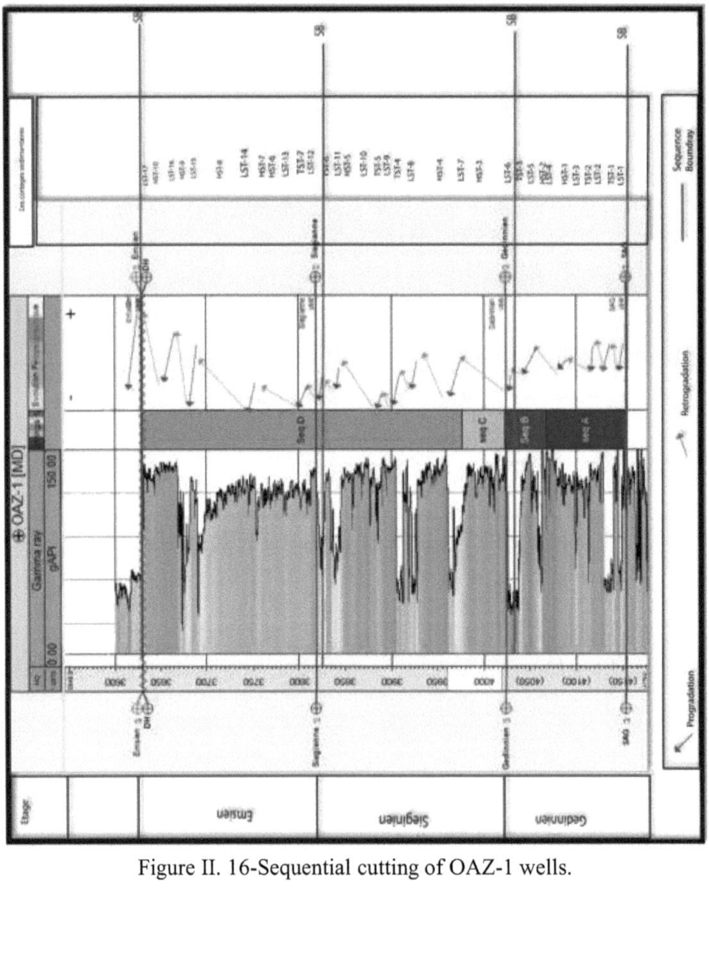

Figure II. 16-Sequential cutting of OAZ-1 wells.

Sequence stratigraphy of the MAME-1 well

Four sequences have been defined:

Sequence A

47.96 m thick, characterized by two sedimentary processes, LST-1 and HST-1 (Figure II.17)

Sequence B

With a thickness of 49.46m, characterised by the presence of a single LST-2 sedimentary process.

Sequence C

10.73m thick, consisting of HST-2 sedimentary material.

Sequence D

Made up of alternating low-level marine prisms with high-level prisms and some passages of transgressive processions.

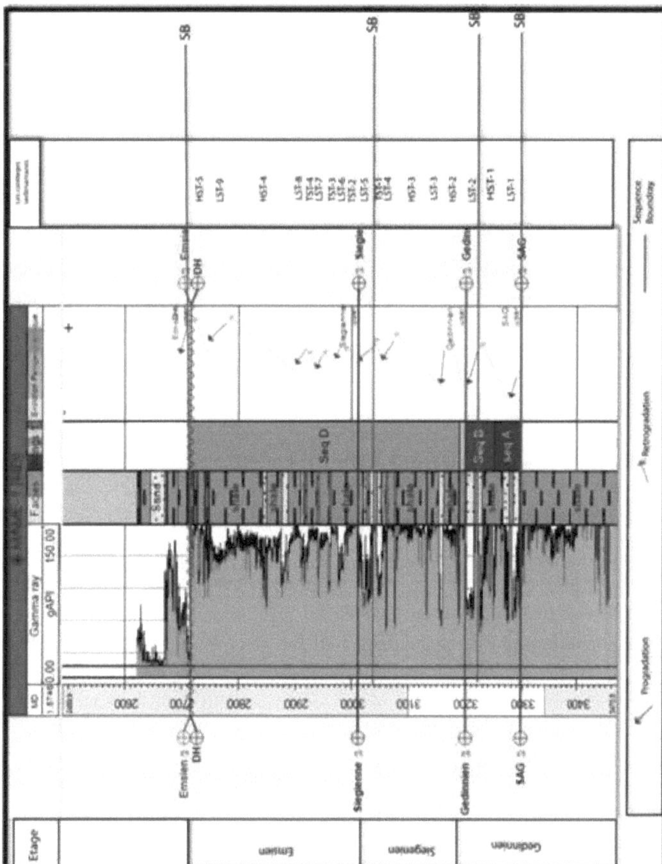

Figure II. 17- Sequential slicing MAME-1 wells

11.9

11.10 Lithostratigraphic correlation

Lithostratigraphic correlation consists of following the sandstone packages laterally (this correlation is typically lithological). The aim is to find the Gedinnian sandstone bars (lower N2 and upper N2) and the Emsian sandstone bar (N1). We established three correlation profiles: (Figure II.18)

Profile A: oriented E-W, it passes through the MAMI-1, OAZ-1 and GEJ-1 wells.

Profile B: oriented E-W, it passes through the MAMS-1, AET-1 and GEJ-1 wells.

Profile C: oriented NW-SE, it passes through the OAZ-1. WMAMI-1, MAMS-1.

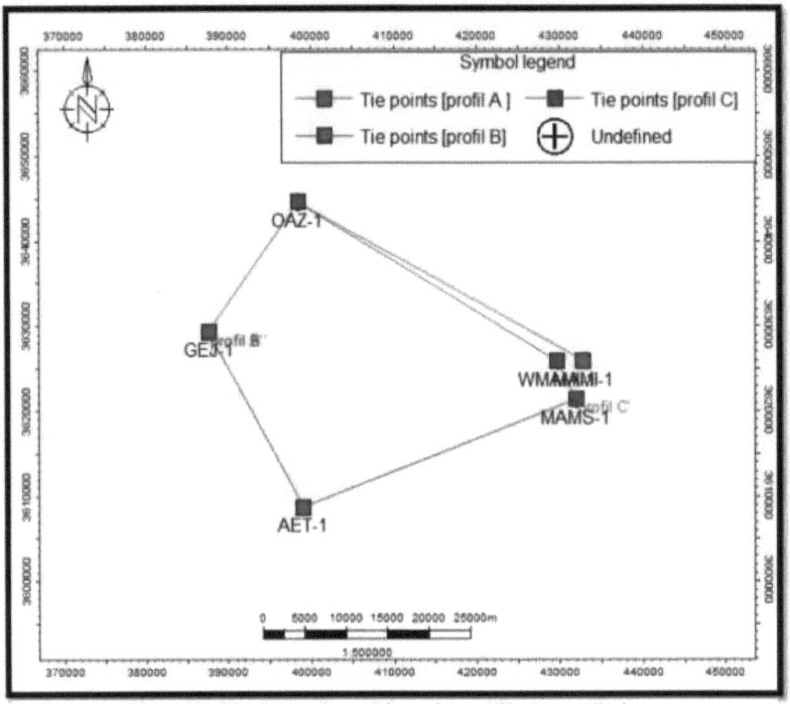

Figure il. 18- Position maps of correlation proms

In this section, log data analysis is the main tool for determining the extension and variation of Lower Devonian facies. Thanks to the regional correlations between the wells, the logs make it possible to follow the evolution of the sequences and their thicknesses from one well to the next.

From the lithostratigraphic correlation shown in Figure II.18 and Figure II.19 and Figure II.20 we can see :

In profile A: (Figure II.19)

Sequence A sandstone levels are found in wells MAMI-1 and OAZ-1 and are absent in GEJ-1.

The two sandstone bars of sequence B only exist in the MAMI-1 well, in the GEJ-1 well we found a single sandstone bar which corresponds to the lower bar. In the OAZ-1 well, the lower sandstone bar becomes .

In profile B: (Figure II.20)

The A-sequence gravelly levels present in the MAMS-1 well thinned out in the AET-1 well, and were absent in the GEJ-1 well.

The N1 sandstone bar is found in the MAMS-1 and GEJ-1 wells, but in the AET-1 well it is thinner than in the other two wells.

In profile C: (Figure II.21)

The N1 sandstone bar is thinner in the MAMS-1 well than in the OAZ-1 well.

In the OAZ-1 well, the lower thick bar tapers towards the MAMS-1 well.

11.11 Conclusion

Based on the log and pit data from the MAMI-1, OAZ-1, AET-1, GEJ-1 and MAME-1 wells, we established a sequence breakdown by correlating it with the Beicip section. In all the wells, the Lower Devonian presents four (4) genetic sequences. In the MAMI-1 and OAZ-1 wells, the four sequences A, B, C and D are thicker. On the other hand, in the AET-1 and MAME-1 wells, the four sequences are moderately thick if we compare them with the two previous wells. In well GEJ-1, the four sequences are . Sedimentological analysis (sequence stratigraphy, regional correlations) has enabled us to identify reservoirs in the Gedinnian, Siégenian and Emsian, corresponding to sandstone complexes in marine environments.

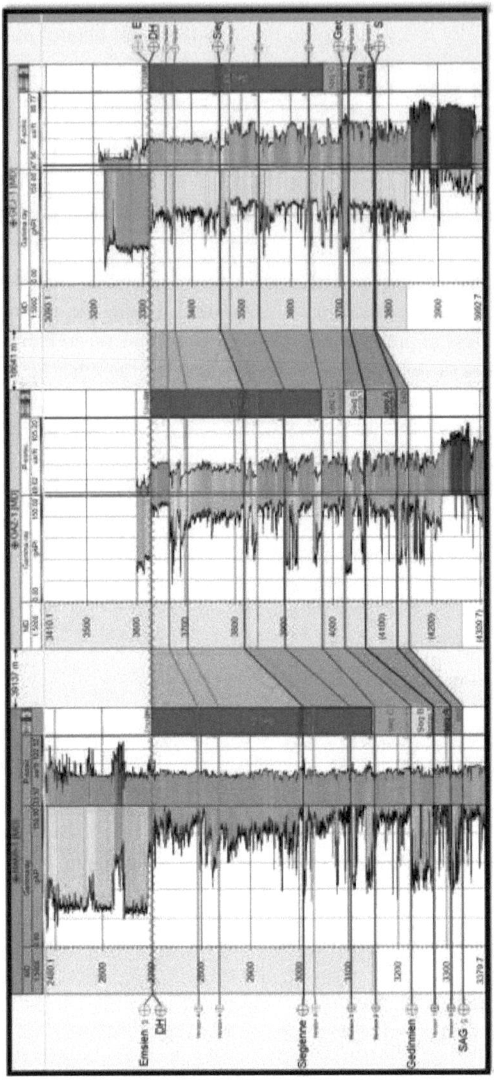

Figure II. 19- Correlation of the wells studied with the MAMI-1 reference well

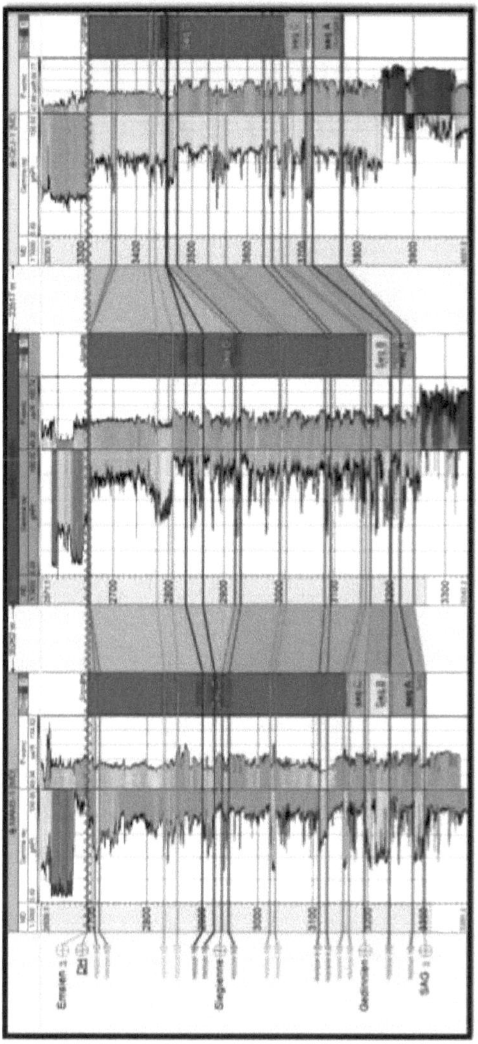

Figure II. 20- Correlation logs for the MAMS-1, AET-1 and GEJ-1 wells.

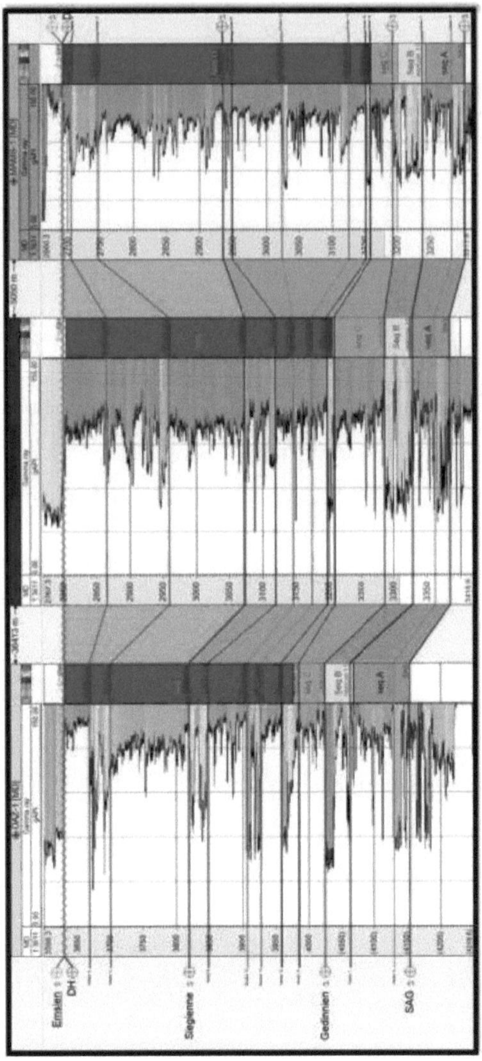

Figure II. 21- Log correlation across wells OAZ-1, WMAMI-1, MAMS-1.

11.12 Interpretation of isobath maps

The isobath maps show the depths of the roofs of the various Lower Devonian reservoirs (Gedinnian, Segentian and Emsian). These maps also allow us to follow the lateral variations in the depths of the Lower Devonian in the region studied.

a) The Gedinnian

The isobath curves show an increase in depth towards the north-west, while the south-east shows shallow depths reaching 2400m (Figure II.22).

b) Le Siégenien

Analysis of the map shows that the depth of the Siégenian roof increases to the north-west with a depth reaching 3000m, the shallow depth is marked at the MAMS-1 well (Figure II.23).

c) The Emsian

Figure II.24 shows the depth of Emsian roof, which is greatest at the OAZ-1 well, where it reaches a depth of 2,800m. Moving south-eastwards, the depth decreases to 1980m.

II.11.1 Conclusion

Analysis of the isobath maps shows that there is an overall deepening towards the north-west. Depth decreases towards the south-east, and subsidence is more significant in the north-west than in the south-east, and it can be seen that the roofs of the Emsian, Segentanian and Gedinnian are parallel, as the isobath maps have an almost identical appearance.

Gedinnian isobath map

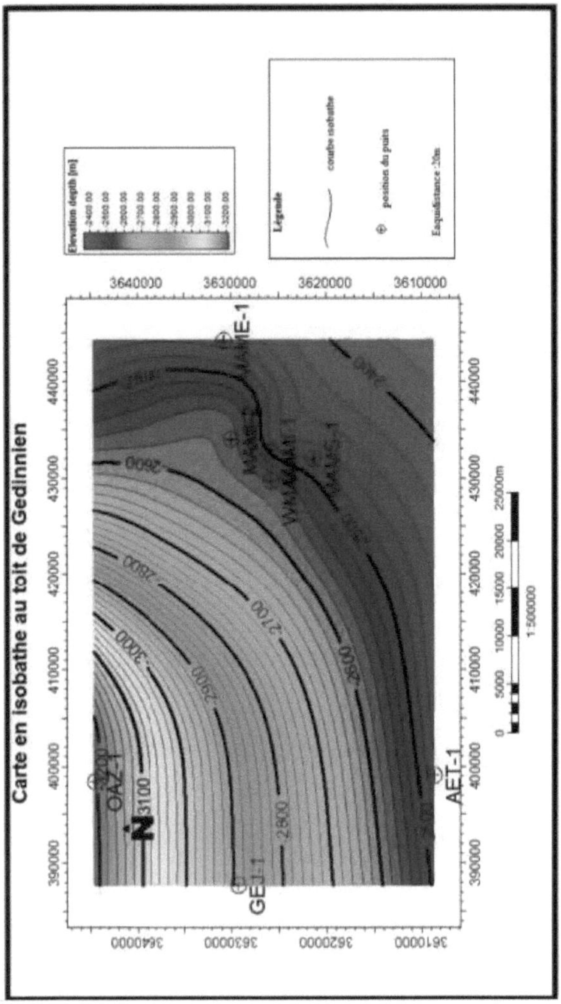

Figure II. 22- Isobath map of the Gedinnian roof

Siégenien roof isobath map

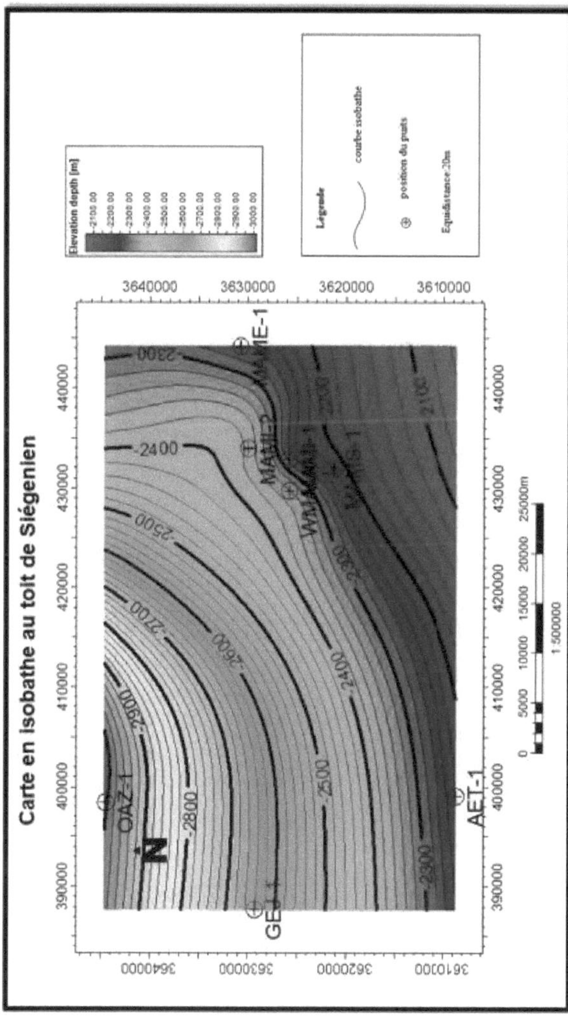

Figure II. 23- Isobath map at the Siégenien roof.

Emsien roof isobath map

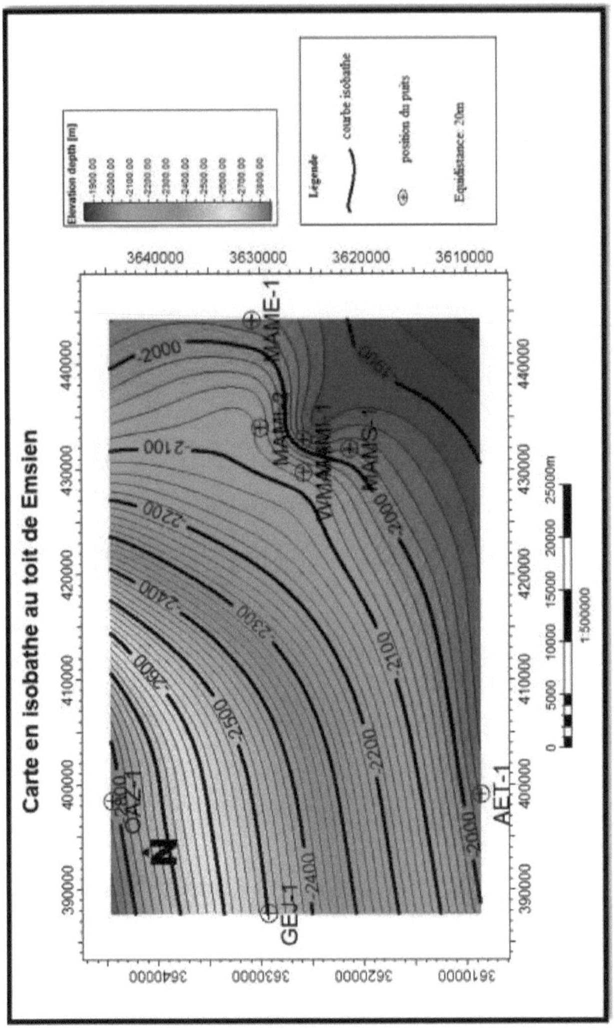

Figure II. 24- Isobath map of the Emsian roof

Chapter III

Petro-physical analysis analysis and statistical study.

111.1 Introduction

The objective a petrophysical study is to evaluate the petrophysical characteristics (porosity, permeability) across the reservoirs studied. Each reservoir must be characterised by its parameters such as porosity, permeability, thickness, oil, gas and water saturation. Of all the parameters already mentioned, the most important to study are porosity and permeability. To better characterise the study area, we analysed cores (horizontal samples), based on the petrophysical parameters (porosity and permeability):

- □ Depth versus porosity diagrams.
- □ Depth versus permeability diagrams.
- □ Making cross-plots (K/0).

Characterisation of reservoirs

The diagram (Figure III.1) shows the typology of reservoirs as a function of porosity and permeability values, and we have based our interpretation on this diagram.

111.2 Graphical representation

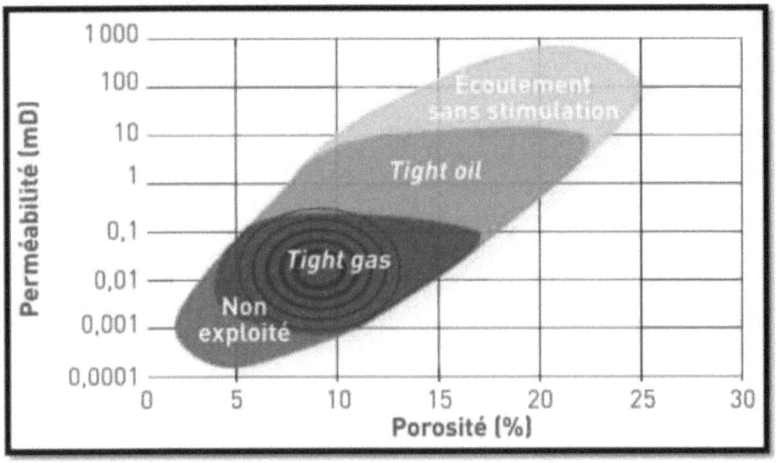

Figure III. 1- Typology of reservoirs according to porosity and permeability (Total, 2011)

The study of the horizontal evolution of porosity and permeability was carried out on four wells:

a) The Gedinnian period

The GEJ-1 well

The distribution of porosity as a function of depth, as shown in Figure III.2, can be subdivided into three intervals:

From 3814m to 3806m: In this interval the porosity varies between 0% and 3%.

From 3800 m to 3798 m: porosity varies between 0% and 2%.

From 3798 m to 3796 m: porosity varies between 6% and 9%.

The distribution of permeability as a function of depth is shown in the diagram in Figure III.1. Very low values of porosity are found at depths of 3814m. The average value of permeability is between 0.003mD and 0.011mD, which corresponds to the interval 3809m-3806m, while the maximum value of permeability is found between depths of 3798m and 3796m.

The OAZ-1 well

The distribution of porosity and permeability as a function of depth, these two parameters have almost identical behaviour: low porosity values are found at depths of 4146.5m and 4136m and low permeability values are found at depths of 4136m and 4146m. (figure III.2)

The MAMI-2 well

The diagram in Figure III.3 shows the distribution of permeability as a function of depth. The curve is almost linear, with values varying between 7 and 10 mD. Four intervals can be distinguished for porosity:

From 3352m-3344m: porosity varies between 3 and 14%.

From 3344m-3342m: porosity is 18.9%.

From 3342m to 3338m: porosity varies between 4% and 15%.

From 3336m-3335m: porosity is between 11% and 16%.

The MAMS-1 well: The diagram in Figure III.3 shows the distribution of permeability and porosity as a function of depth. Low porosity values are found between 3240 and 3234m, high values between 3234m and 3230m and medium values between 3230m and 3224m**.** The distribution of permeability as a function of depth shows that permeability deteriorates with depth.

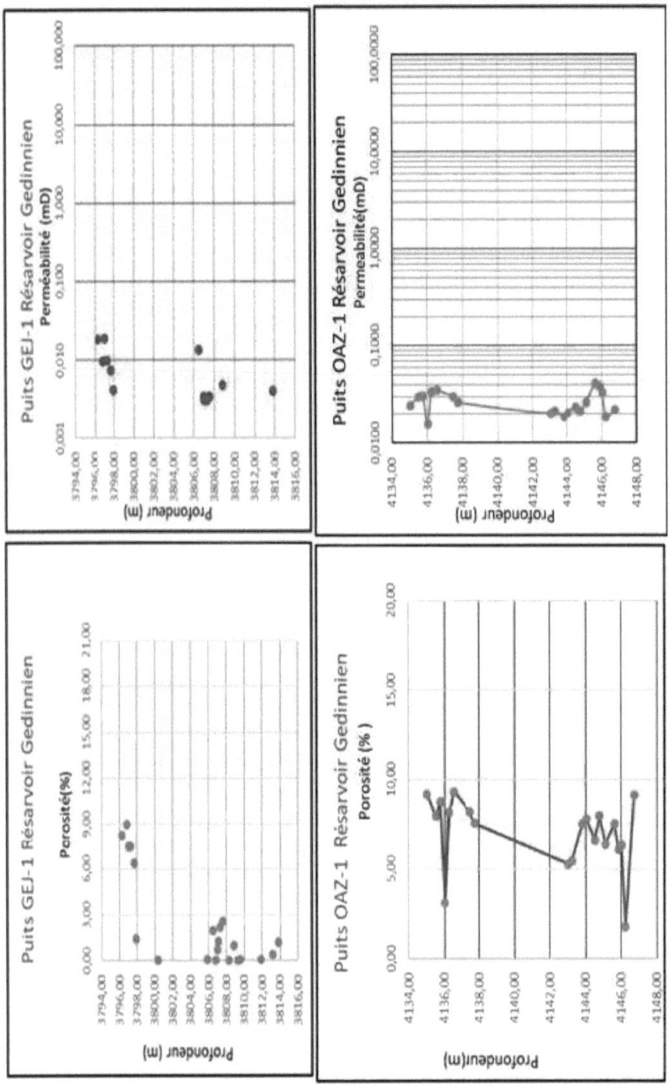

Figure III. 2- Distribution of porosity/permeability as a function of depth in the Gedinnian reservoir.

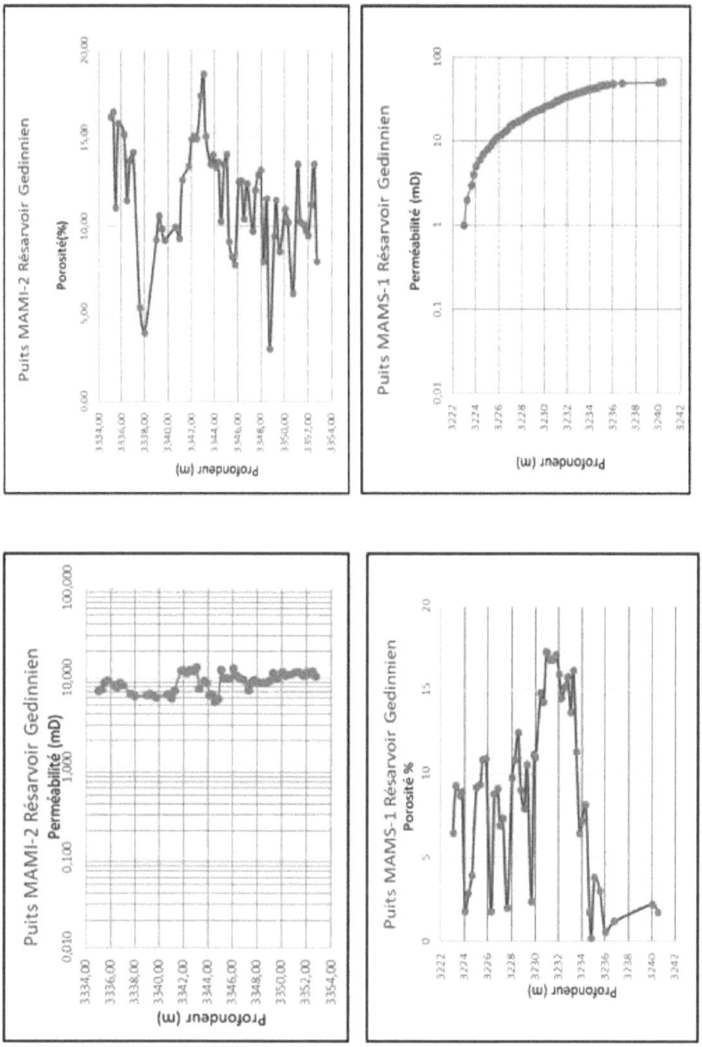

Figure III. 3- Distribution of porosity/permeability as function of depth in the Gedinnian reservoir.

The OAZ-1 well

On the cross plot (figure III.4), we can see that the repair of the points shows two zones: the first is marked by negligible permeability [0.01 ; 0.1 mD] and very low porosity [2 ; 3 %], the second zone is formed by very low permeability [0.01 ; 0.1 mD] and low porosity [5 ; 10 %], reflecting an unconventional compact reservoir (Tight gas).

The GEJ-1 well

On the cross plot (figure III.4), we can see that the distribution of points highlights two zones: the first zone characterised by very low permeability [0.01; 0.1 mD] and negligible porosity [2; 3%], and the second zone characterised by very low permeability [0.01 ; 0.1 mD] and low porosity [9; 10%], forming an unconventional reservoir (Tight gas).

The MAMS-1 well

On the cross-plot (Figure III.4), the distribution of points makes it possible to distinguish three zones: the first is characterised by very low permeability values [0.01; 0.1 mD] and low porosity [0.5 to 4%], which implies that this is a compact reservoir that has not been exploited; the second is represented by permeabilities [0.01; 0.1 mD] and porosity [6 ; 11%] forming an unconventional compact reservoir (Tghit gas). The third is formed by porosities [12 ; 17 %] and permeability [0,1 ; 2mD] constituting an unconventional compact reservoir (Tghit oil).

The MAMI-2 well

On the cross-plot in Figure III.4, the distribution of points shows two different zones: the first zone is characterised by a low permeability value [>10mD] and porosity
[The second zone is characterised by a permeability [>10mD] and porosity [8 to 19%] forming a fissured reservoir.

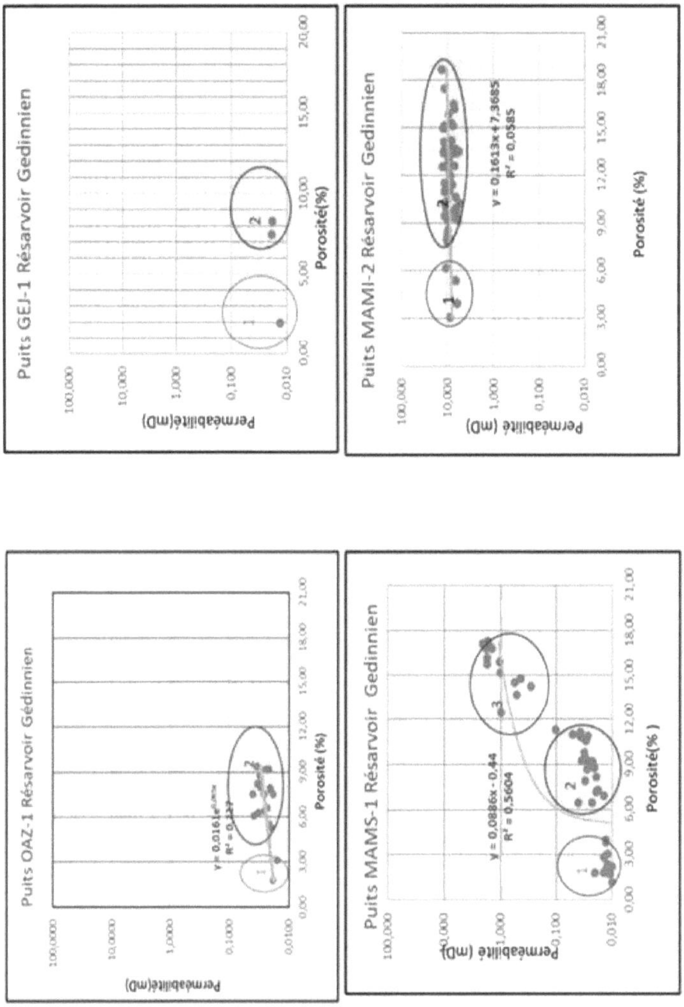

Figure III. 4- Distribution of porosity as a function of permeability in the Gedinnian reservoir.

b) Le Siégenien

The MAME-1 well

The diagram in Figure III-5 shows a distribution of permeability values as a function of depth. It can be seen that the permeability values are around 0.010 mD and <100 mD. A peak in the maximum permeability value is recorded at a depth of 3021m.

The diagram in Figure III-4 shows that porosity at intervals of 3038m and 3020m varies horizontally between 3 and 12%.

The MAMI-2 well

The diagram (Figure III.5) of porosity as a function of depth is the same as that of permeability as a function of depth, in which four intervals can be distinguished.

The porosity values vary between 0% and 10% and the permeability values between 1 and

80mD.

The MAMI-1 well

The diagram (Figure III.6) of porosity as a function of depth shows that the low values are found at depths of 3112m and 3110m, and the high values at depths of 3110m and 3105m.

The permeability versus depth diagram shows a low value of 0.04mD at interval [3110m-3109m], the high permeability value is 1.1 mD found at interval [3106m-3105m].

The GEJ-1 well

The diagram in Figure III-6 shows a distribution of permeability values as a function of depth. Three intervals can be distinguished:

From 3554m to 3546m: the permeability value is between 0.001 and 0.010 mD.

From 3540m to 3536m: permeability value between 0.001 and 0.010 mD.

At a depth of 3538 m: the permeability is 1mD.

The porosity versus depth diagram shows three zones:

From 3556m to 3546m: the porosity value varies between 5% and 10%.

From 3546 m to 3542 m: the porosity value varies between 0% and 4%.

From 3540m to 3536: the porosity value varies between 0% and 5%.

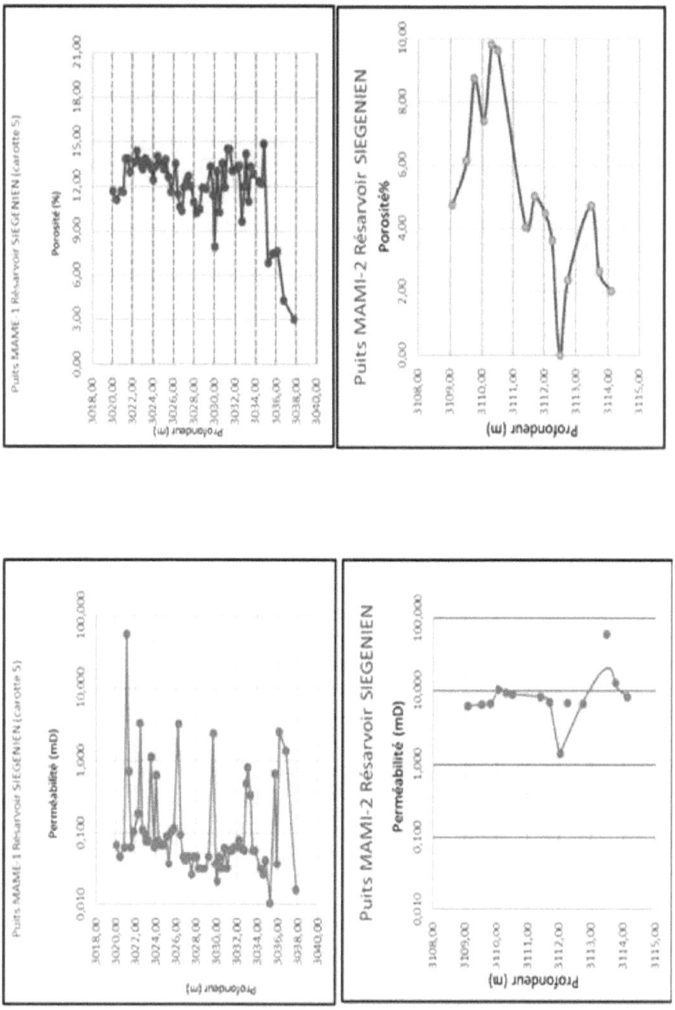

Figure III. 5- Distribution of porosity/permeability as a function of depth in the Siégenien reservoir.

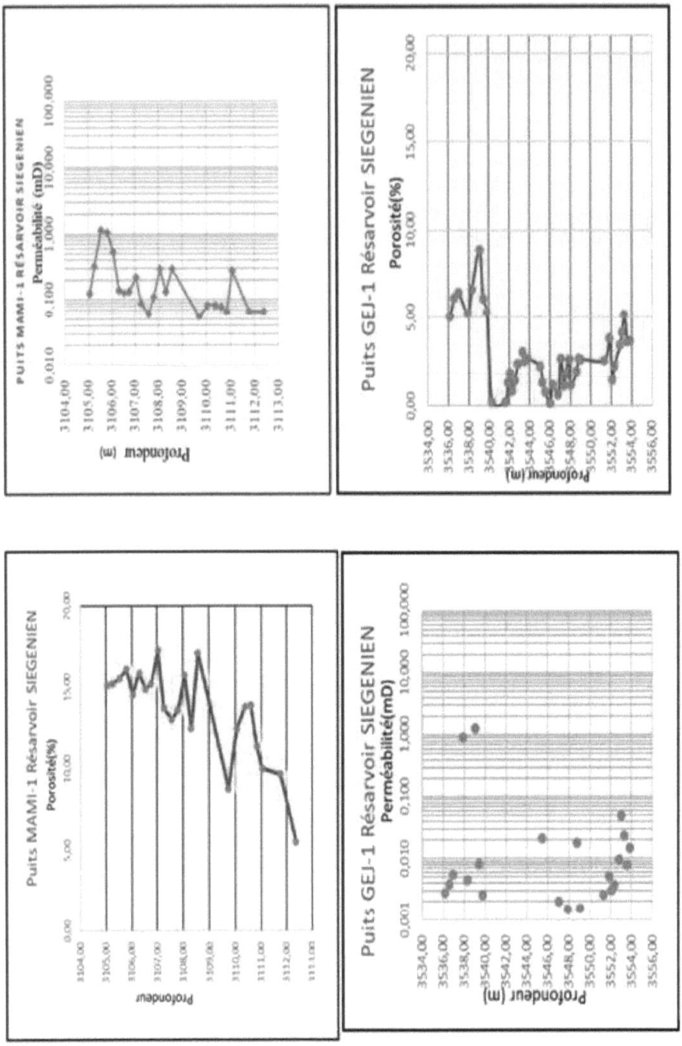

Figure III. 6- Distribution of porosity/permeability as function of depth in the Siégenien reservoir.

The GEJ-1 well

On the cross-plot in Figure III.7, the points are mainly concentrated in two zones: the first is characterised by permeabilities [0.01 to 0.1 mD] and porosity [0% to 5%], while the second is characterised by permeability [1 to 10 mD] and porosity [5% to 9%], suggesting a non-conventional reservoir (Tghit oil).

The MAMI-2 well

On the cross-plot in Figure III.7, the points show two zones: The first zone is characterised by a permeability [1mD] and porosity [5%] which is synonymous with a non-conventional reservoir (Tghit oil). The second is characterised by low permeability [10 and 100mD] and low porosity [2 to 9%], which indicate a conventional reservoir.

The MAMI-1 well

On the cross-plot in Figure III.7, the points show four zones: with permeability values of around 0.1 and 1.00 mD and porosity values that differ, the first zone corresponds to an unconventional reservoir (Tghit gas). The second zone and third and fourth zones correspond to unconventional reservoirs (Tghit oil).

The MAME-1 well

The cross-plot in figure III.7, shows a distribution of points in eight zones: the first three zones have permeability values of around [0.01 and 1mD] and porosity values which break down: the first zone has a porosity of [0% to 5%] which indicates an unexploited reservoir, the second [7% to 8%] is an unconventional reservoir (Tghit gas).The fourth and fifth zones form unconventional reservoirs (Tghit oil), as do the sixth and seventh zones. Zone eight has a permeability of between 10mD and 100mD, forming a conventional reservoir.

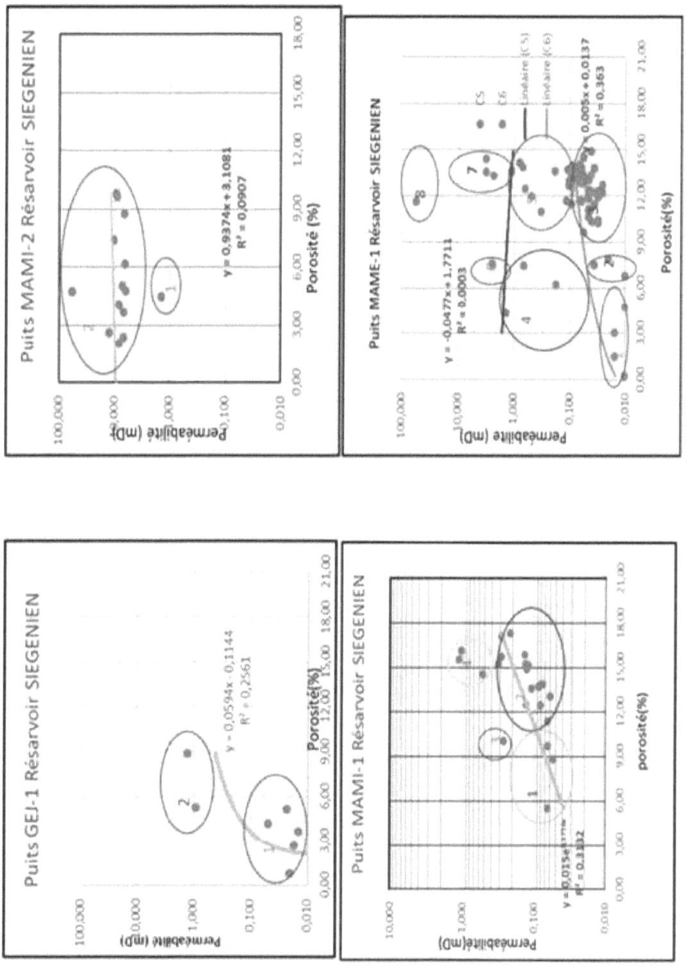

Figure III. 7- Distribution of porosity as a function of permeability in the Siégenien reservoir

c) The Emsian

The MAMI-1 well

The porosity versus depth diagram in Figure III-8 shows values between 4% and 16%, while the permeability diagram shows values between 0.04 and 10 mD.

The MAMI-2 well

The diagram (figure III.8) of permeability versus depth is the same as that of porosity versus depth. The maximum value for porosity is recorded at depths of 2866m and 2864m, whereas the maximum value for permeability is recorded at depth 2866m.

The OAZ-1 well

The diagram (Figure III.9) of porosity versus depth and permeability versus depth shows that the porosity values are around 0% and 1% and the permeability values are between 0.001 mD and 10 mD.

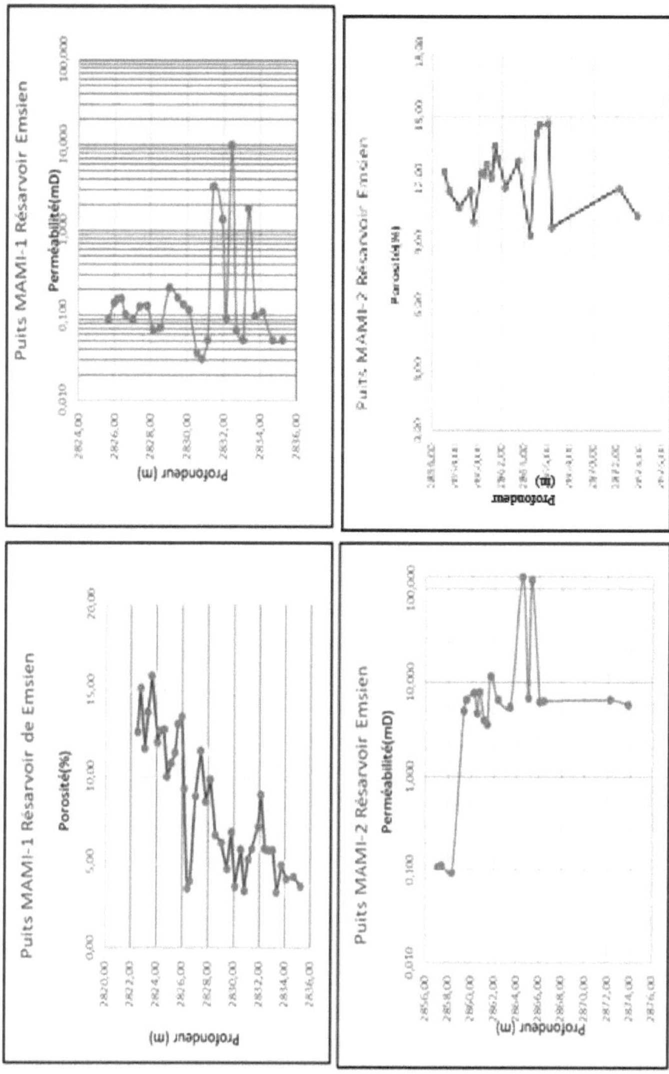

Figure III. 8- Distribution of porosity/permeability as function of depth in the Emsian reservoir

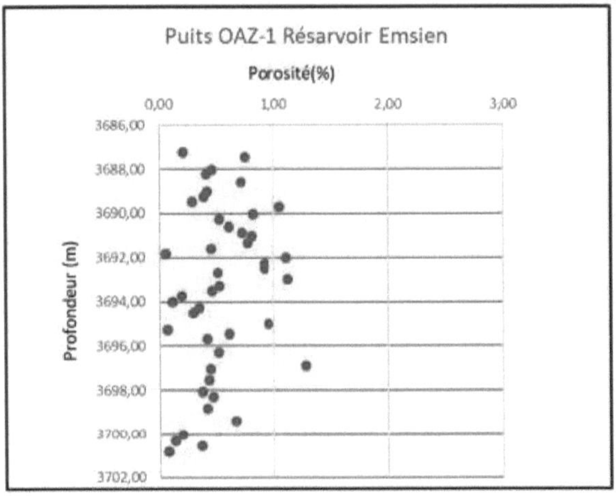

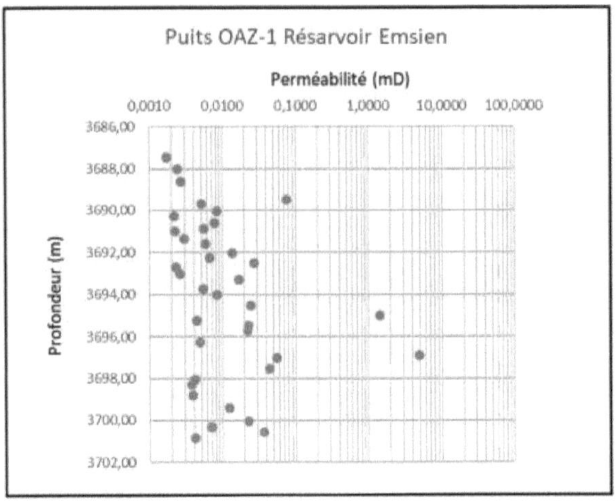

Figure III. 9- Distribution of porosity/permeability as a function of depth in the depth in the Siégenien reservoir.

a. The Emsian

The OAZ-1 well

The cross-plot in figure III.10 shows two zones: the first zone is marked by very low permeability values [0.01 to 0.1 mD] and porosity [0 to 2%], thus forming an unexploited reservoir.the second zone has permeability values ranging from 1 to 10.0 mD and porosity from 0 to 3%.

The MAMI-1 well

The cross-plot in figure III.10, The first zone has very low permeability [<0.1 mD] and porosity [3 to 6%], indicating a non-conventional reservoir (Tghit gas).The second zone is characterised by a value of permeability [0.1 to 1 mD] and porosity of 9 to 15%, corresponding to an unconventional reservoir (Tghit oil). The third, fourth and fifth zones have values of low permeability [0.1 to 10 mD] and porosity [6 to 11%], making it an unconventional reservoir (Tghit oil).

The MAMI-2 well

On the cross-plot in Figure III.10, the distribution of points shows three different zones: The first zone is characterised by low permeability [0.09; 0.1mD] and porosity [11% to 13%] forming a compact unconventional reservoir (Tghit gas). The second zone shows low values of permeability [1 ;10 mD] and porosity [9 to14%], reflecting the presence of an unconventional reservoir (Tghit oil).

III.3 Conclusion

Cross plot analysis shows that :

The petrophysical properties of the Gedinnian, Seigenian and Emsian reservoirs are very low, which explains the tight nature of the reservoir.

The correlation coefficients (2) are close to zero, which explains the poor quality of these reservoirs due to the heterogeneity of the environment.

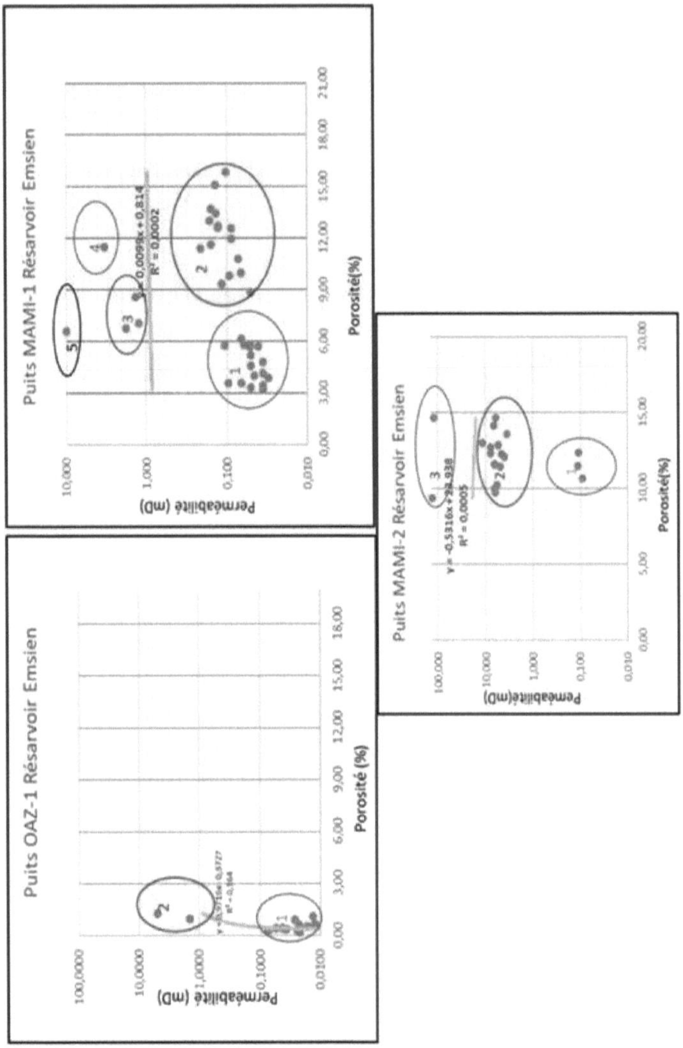

Figure III. 10- Distribution of porosity/permeability as function of depth in the Siégenien reservoir.

III.4 Reminder of the statistical method

1. **Arithmetic mean** :

The average is a positional parameter. The average is obtained by taking the sum of the values divided by the number of values.

— 1+ 2+···

The variance is equal to the sum of the squares of the deviations from the mean divided by the total number of people. If x = (x1, x2, ...xn). The variance is denoted var (x), 2, or 2.It is represented by the following equation: $2(X) = \sum_1 (-)2$ With N - total number of employees; $\bar{}$ - arithmetic mean and xi - variable.

3. Standard deviation

$$S(X) = \sqrt{2(\ )}$$

This is the square root of the variance

III.5 Statistical results

a) The Gedinnian period

Table III.1 shows the porosity and permeability values for the Gedinnian reservoir. Maximum porosity values are recorded at the MAMI-2 well and also for permeability.

Well	X(m)	Y(m)	Average permeability (md)	Average Porosity (%)	Floor from tank
OAZ-1	398428.70	3644563.40	0,03	7	Gedinian
GEJ-1	387711.19	3629312.03	0,01	2,46	Gedinian
MAMS-1	431952.49	3621359.63	0,34	8,93	Gedinian
MAME-1	444182.20	3630816.20	2,53	6,37	Gedinian
MAMI-2	433944.60	3630034.00	9,23	11,57	Gedinian

Table III. 1- Porosity and permeability of the Gedinnian reservoir

b) Le Siégenien

Table III.2 shows the porosity and permeability values for the Siégenien, porosity values vary between 1.45% and 13.51%, while permeability values vary between 0.24 mD and 11.37 mD. The maximum porosity value is found in well MAM-1 and the maximum permeability value in well MAMI-2.

Well	X(m)	Y(m)	Average permeability (md)	Average Porosity (%)	Tank level
MAMI-1	432829.09	3625900.75	0,24	13,51	Siégenien
GEJ-1	387711.19	3629312.03	0,11	1,45	Siégenien
MAME-1	444182.20	3630816.20	1,20	11,89	Siégenien
MAMI-2	433944.60	3630034.00	11,37	5,38	Siégenien

Table III. *2*- Porosity and permeability values for the Siégenien reservoir.

c) The Emsian

Table III.3 shows the variations in the porosity and permeability of the reservoir. Emsian. There are appreciable porosity and permeability values in the wells MAMI-2.

Well	X(m)	Y(m)	Average permeability (md)	Average Porosity (%)	Floor tank
MAMI-1	432829.09	3625900.75	0,90	8,22	Emsien
OAZ-1	398428.70	3644563.40	0,11	0,54	Emsien
MAME-1	444182.20	3630816.20	0,05	7,29	Emsien
MAMI-2	433944.60	3630034.00	17,54	12,02	Emsien

Table III. 3- Porosity and permeability values for the L'Emsien reservoir.

Statistical study on all the wells (table III.4) informs us that the maximum porosity variability is recorded at well MAMI-1 (2=13.99%) with a standard deviation of (S = 2.65 %).and maximum variability is found at well MAMI-2 (2=13.99%) and standard deviation (S =0.22).

Well	X(m)	Y(m)	Var. (2)of permeability	Var. (2)of Porosity	Standard deviation of Permeability	Standard deviation Porosities
MAMI-1	432829.09	3625900.75	0.22	13.99	0.33	2.65
OAZ-1	398428.70	3644563.40	Var <0	20.87	0.04	3.23
GEJ-1	387711.19	3629312.03	0	0.51	0.05	0.50
MAMS-1	431952.49	3621359.63	/	/	/	/
MAME-1	444182.20	3630816.20	2.66	14.31	1.41	3.28
MAMI-2	433944.60	3630034.00	18.61	13.77	3.52	3.03

Table III. 4 - Porosity and permeability statistics for the six wells.

III.6 Reminder on geostatistics

111.6.1 Introduction

Geostatistics is a physical science (Chauvet, 1999), study of natural phenomena spread over space (regionalised phenomena) and/or time.

Geostatistics, or the theory of regionalised variables, is concerned with the study of random functions. Mathéron (1962a and 1962b) introduced the notion of regionalised variables to quantify a random phenomenon with a spatial variability structure. He provided a theoretical basis for the probabilistic analysis of regionalised variables. Geostatistics has already been used in a number of research projects in geology and applied geology (Chiasson et al., 1995; Bolle, 2000; Marache et al., 2002; khlalfaoui. S,2008).

111.6.2 Geostatistical tools

The variogram

The basic tool that characterises the random function is the variogram. It is generally used to model the way in which two values in space or time are correlated. Mathematically, it is the difference between two measurements or the squared deviations of the values.

Z(x) and Z(x+h) observed at points x and x+h: { $\gamma(h) = (1/2)\, E\{[Z(x) \square Z(x+h)|-\}$.(Figure III.11)

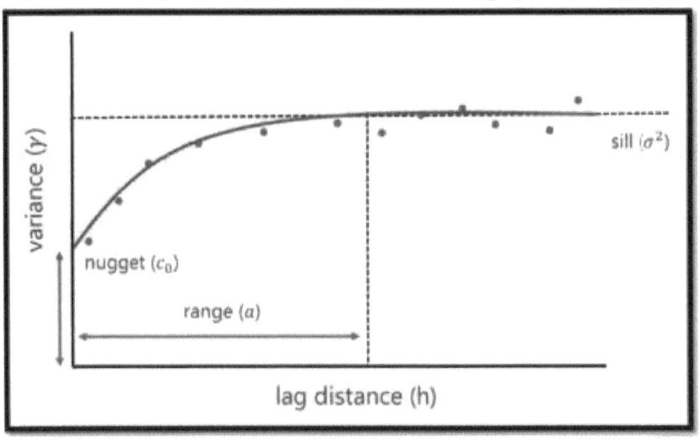

Figure III. 11- Graphical representation of a variogram.

Nugget effect:) : Discontinuity in the variogram at a distance of less than data spacing or zero separation distance. The larger the Nugget, the more discontinuous the data.

Sill (level C): the point at which the behaviour of a variogram changes, from which point the link between measurements decreases. The greater the sill, the weaker the correlation and the more unreliable the estimate.

Range: the distance beyond which two measurements are not correlated.

111.6.3 Common variograms

The geostatistical literature offers numerous functions that satisfy the properties of the semi-variogram as presented below. These parameterised functions must be able to describe the various components (range, bearing, nugget). They must also manage the behaviour of the function at the origin (linear trend, horizontal or vertical tangency, etc.). We will only present examples of variogram models here. (Table III.5)

Model	Expression	Representation
Nugget effect	$\gamma(h)=0$ h=0 $\gamma(h)=C$ $\|h\|>0$	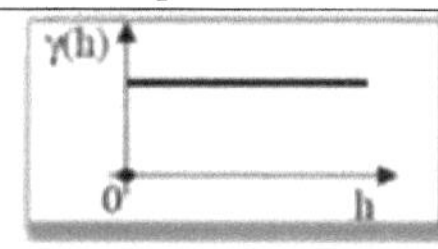
Spherical	$\gamma(h) = 0 + [2\,3(\frac{h}{a}) - 1\,(\frac{h}{a})^3]$ $\|h\| < a$ et $\|h\| > a$.	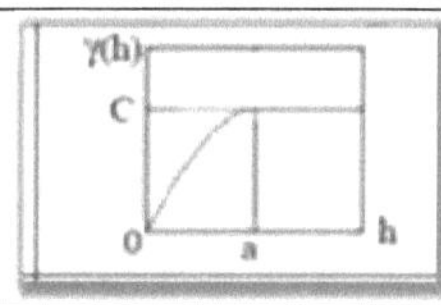
Exponential	$\gamma(h) = 0+C\,[1-\mathrm{Exp}\,[\frac{h}{a}]]$.	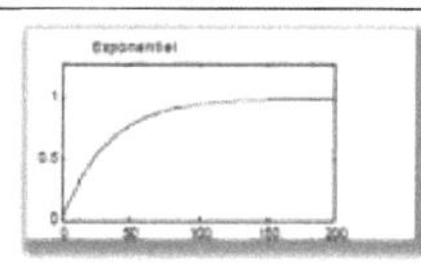
Gaussian	$\gamma(h) = 0+C\,[1-\mathrm{Exp}\,(-\frac{\|h\|^2}{a^2})$.	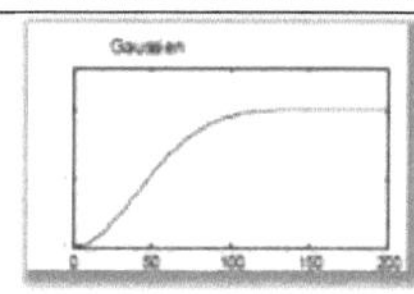
Power	$\gamma(h)=\|h\|^{\alpha}$ a=1 linear. a=2 parabolic	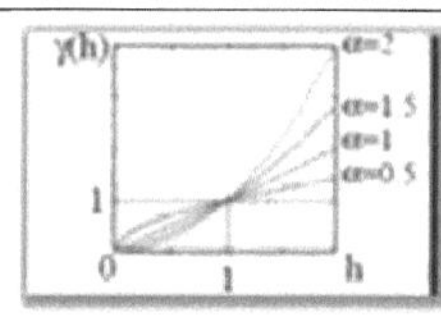
Logarithmic	$\gamma(h)=C.\log\frac{h}{a}$ $\|h\|>a$	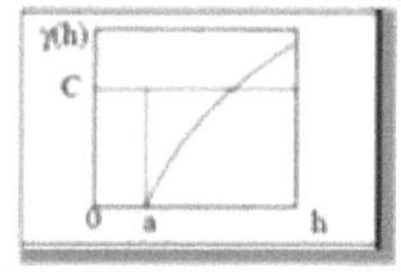
Cardinal sinus	$\gamma(h)=C\,[1- \quad]$ $r=\frac{h}{a}$	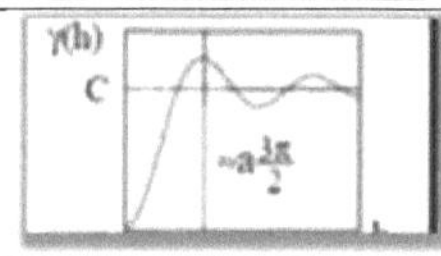

Table III. 5- Examples of variogram models.

III.6.4 Graphical representation of a variogram

Under the so-called "intrinsic" assumption that the semi-variogram depends only on the vector h, this theoretical variogram can be estimated by a so-called experimental semi-variogram y*(h) which is calculated using the following formula:

$$\gamma^*(h)=\frac{1}{2N(h)}\sum_{i=1}^{N(h)}[Z(x)-Z(x+h)]^2$$

Where N(h) is the number of pairs of measurements Z(x) and Z(x+h), this experimental semi-variogram is then fitted using one of the well-established theoretical models. The experimental variogram can be calculated in 1, 2 or 3 dimensions. The results of the calculation of an experimental variogram can be represented graphically.

III .7 Geostatistical results

a) Surface Vriograms

The surface variogram for permeability (Figure III.12) shows a directional anisotropy of 90° and also for the porosity variogram. (Figure III.13) we can see that these two parameters are really linked.

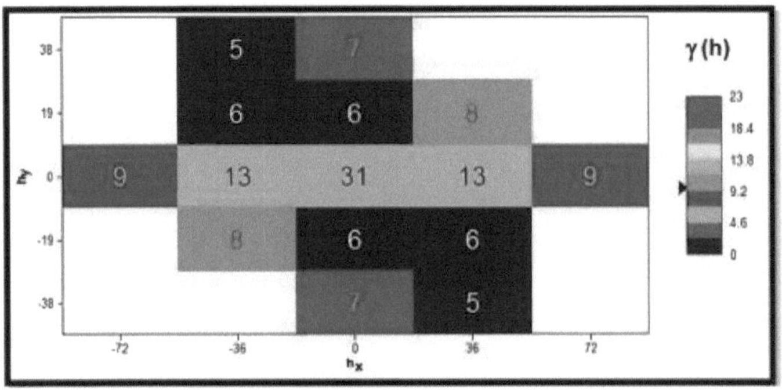

Figure III. 12 - Variogram of permeability surface.

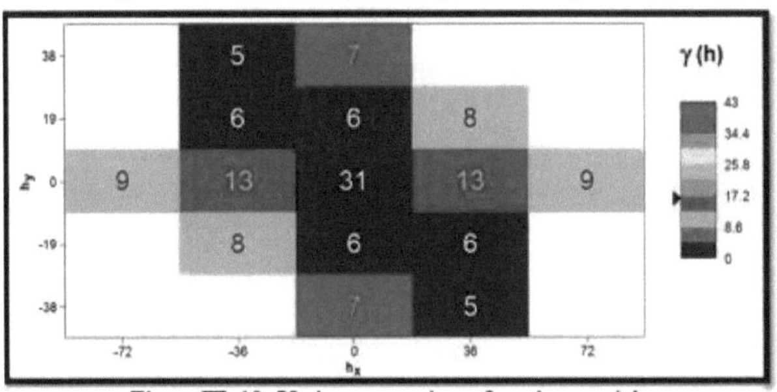

The geostatistical variography (figure III.14) shows a relatively homogeneous structure for the two parameters (table III.6) porosity and permeability. The variograms of the calculated surfaces indicate an anisotropy towards the 90° direction, and a slightly weaker nugget effect for the permeability model. These results confirm the reliability of the data and the significance of the variogram model. For the variation, the C step is more important for porosity than for permeability, which explains the beginning of the decrease in the covariance between measurements by exceeding the range (a). In conclusion, these results of statistical and geostatistical analysis ; can be used estimate these and other parameters using kriging (simple or indicative).

The variable	Nugget (nugget effect: o)	Sill(bearing :C)	Range (range: a)	Coefficient anisotropy	The model
Porosity	5,44	11,56	62,7	0,82	Gaussian
Permeability	1,45	8,99	42,18	0,97	Gaussian

Table III. 6 - Comparison between two variables (Porosity and Permeability).

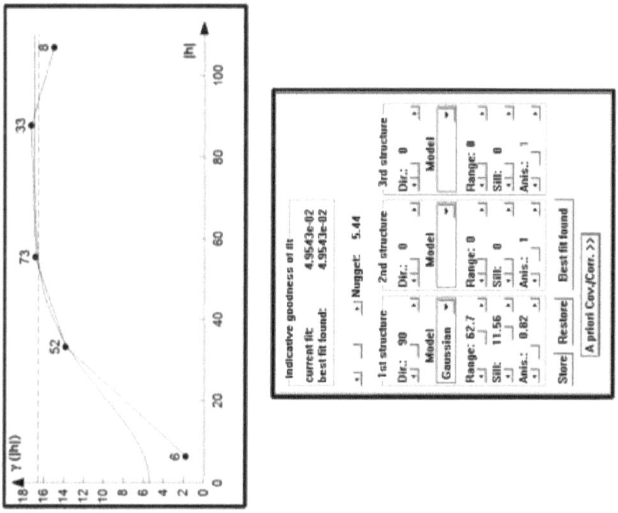

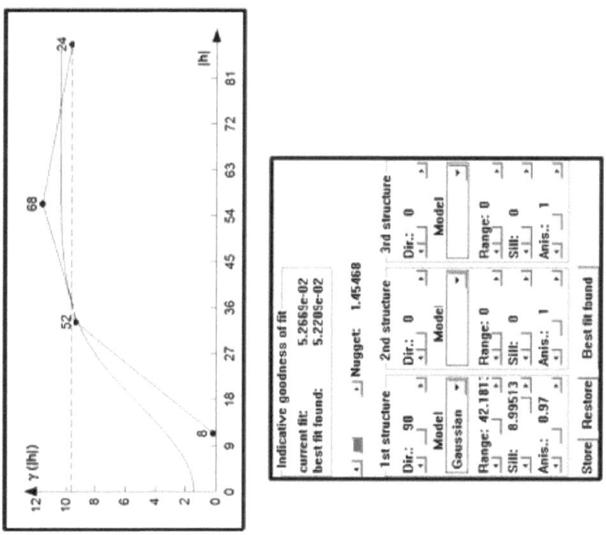

Figure III. 14- Presentation permeability and porosity variograms (left: permeability variogram, right: porosity variogram)

Chapter IV

Oil potential

IV .1 Oil system

The eastern part of the Benoud sillón contains the Palaeozoic oil plays. The Lower Devonian and Ordovician are the main targets. The Jurassic and Triassic are also potential targets in the northern part of blocks 103a and 312 (for confidentiality reasons, the map of the blocks is not shown).

V V.1.1 Source rocks

In the El Ouabed II perimeter (the area studied), several source rock levels have been identified. Their geochemical characteristics appear to be suitable for hydrocarbon generation. These potential levels, rich in organic matter, are recognised in the Silurian, Devonian, Ordovician and Jurassic.

a. Silurian bedrock

The Silurian consists of black clays, rich in graptolites (radioactive clays). These series, dated Ludlow and Wenlock, are underlain by Ordovician sandstones in almost all the boreholes studied. Its thickness varies from 100 to 400 m.

The most suitable levels, which are rich in organic matter, are the lower radioactive clays and the upper radioactive clays.

The geochemical study carried out by CRD (1992) showed an excellent richness in organic matter in these levels. The average TOC is of the order of 3.97% (Figure IV-1), with high S2 and oil potentials (S2=34.66 mg HC/g rock and PP=39.60 mg HC/g rock) for the sample at elevation 2651.90m from borehole NL-2.

The MAMI-1 well: The radioactive Silurian shows TOC levels ranging from 0.40% to 7.06%. (Interval ranging from 3410m, moderate S2 values. Tmax is between 417 and 460°C and IAT between 3 and 3.5).

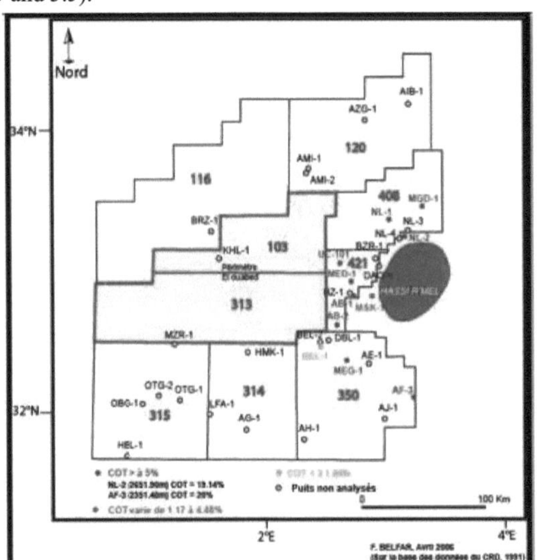

Figure IV. 1 - Map of measurements of C.O.T levels in Silurian parent rock (BELFAR.F, April 2006)

Data from geochemical analyses carried out by BHP (2001) on the lower and upper Radioactive Silurian of a number of wells adjacent to the El Ouabed perimeter show the

following results
can be summarised as follows (Table IV.1):

Well	**TOC (%)**	**S2 (mg HC/g of R)**	**IH (mg HC/g of TOC)**	**Tmax (°C)**	**IAT**
MAMS-1	8,65	7 ,65	116	446 ,75	4 à 5
MAME-1	4,40	5,28	110	450,23	2,8 à 3
MAMI-2	1,52	0,31	31	378,43	3 à 4
AET-1	5,52	6,56	120	440	3 à 3,5
MAMI-1	3,03	5,32	173	447,92	3 à 3,5
OAZ-1	3,99	1,90	56	467,1	3 à 3,5
GEJ-1	4,65	4,95	103	447,1	/

Table IV. 1- Geochemical parameters of Silurian bedrock (BHP, 2001)

Maturation of the Silurian

The Silurian maturation map drawn up on the basis of the available geochemical data and by extrapolation (Figure IV.2) shows 03 distinct maturation zones, following a SE-NW direction from the oil phase in the SE to the gas phase in the NW, passing through the condensate phase in the centre. The tanks are then supplied with gas or condensate around the perimeter.

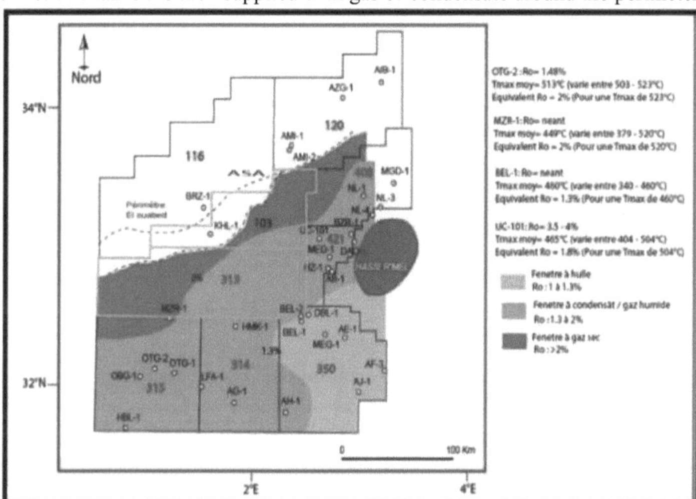

Figure IV. 2- Maturation map of the Silurian bedrock (drawn up by Anadarko (1993)), completed by F. Belfar; Avril, 2007)

b. Devonian bedrock

Only borehole MED-1 in the Gedinnian interval (2632-2760m) has TOC values of up to 1.12%, with S1=1.99 mg HC/g rock and PP=5.76mg HC/g rock.

This presence of free hydrocarbons is also reported in the sandstone levels (Producer Levels) of borehole BEL-1, where the sample at elevation 2251m recorded S1 = 4.09 mg HC /g rock.

Given the lack of organic matter, it is difficult to assess the state of maturation of these series. However, the oil formation zone is found at NL-2 and BEL-1 (pyrolysis T°max of 445°C and I.A.T varying from 1.5 to 3). On the other hand, at MED-1, the T°max are higher (484°C<T°max<537°C and the IAT varies from 3 to 4), corresponding to the oil and condensate-gas zone.

The Frasnian clays are 50 to 70 m thick, generally truncated by the Hercynian unconformity. Well HBL-1 (located to the south of the perimeter) traversed 45 m of grey-black clay attributed to the Frasnian, with a calculated C.O.T. of over 6%, the average S2 peak being 4.3 mg HC/g of rock. These clays are most often found at the end of the oil window, with T°max of 430° to 445°C and an IAT of 3. The organic matter is of type II-III.

c. Ordovician bedrock

The micro conglomeratic clays and the El Gassi clays have a facies consisting of black, carbonated to silty clays, indicating a marine depositional environment.

d. Carboniferous bedrock

The Carboniferous is only present in the Bordj Nili region, which was traversed by boreholes MGD-1 and NL-1. The samples taken and analysed showed a low level of organic matter at NL-1, while it was average at MGD-1 (C.O.T reached 1.15% between 3069.20m and 3106.75m). Rock Eval pyrolysis showed low parameters and a state of maturation in the condensate-moist gas phase (T°max=464°C).

Optical microscopy of the kerogen shows a palynofacies composed of amorphous organic matter with some woody debris, cuticles and spores. The state of maturation of this analysed interval is between the end of the oil phase and the condensate-moist gas phase (I.A. T=2.5 to 3).

IV.1.2 Reservoir rocks

The main potential reservoirs in the region are located in the Palaeozoic:

a. The Devonian reservoir

The reservoir properties of Lower Devonian sandstones are highly variable, ranging from 2 to 30% for porosities and an average of 0.7 mD. for permeability. These reservoirs remain tight. Nevertheless, the Gedinnian reservoir revealed 7.7m3/h of oil in borehole MAMI-1 and the presence of oil in boreholes MAMI-2, MAME-1 and AET-1.

b. The Ordovician reservoir

The M'kratta slab has revealed very good indications, in particular the presence of oil in the Mammi and Arich Et Tergui structures. This reservoir has appreciable porosity, but low permeability (Φ = 1-8% and k = 0.01mD). These poor petrophysical properties are mainly due to the destruction of the primary porosity by diagenetic transformations, particularly siliceous cementation. These properties are sometimes improved locally by the presence of fracturing.

c. The Carboniferous reservoir

The Muscovite sandstones and carbonates that produced oil (8m3/h) at Nili-2 may be prospective to the NE (block 103) and SW (block 312) of the EL OUABED II perimeter. At NI- 2, the porosity is 20% and the permeability 800mD. The Visean, which was detected by seismic in reef form, may also be prospective to the NE and SW of the perimeter.

IV.1.3 Cover rock

The Paleozoic reservoirs are covered by Ordovician, Silurian, Devonian and Carboniferous clays and the Saliferous Triassic.

IV.1.4 Pitfalls

A variety of traps can be explored in the El Ouabed II perimeter, including simple or complex structural traps, mixed traps and lithological traps.

In conclusion, the Benoud Trench is a transition zone between the Saharan Platform to the south, which has been stable since the Palaeozoic era, and the active Meso-Cenozoic Saharan Atlas to the north. The position occupied by the Benoud Trench reflects the effects of the different tectonic phases of the Hercynian and Alpine cycles.

General conclusion

The Lower Devonian in the study area is made up of deposits that are developed throughout the Algerian Sahara, with a sandy clayey character, and are found in three stages, the Gedinnian, Seigenian and Emsian. This study is based on the description of cores from five wells (MAMI-1, MAME-1, GEJ-1, OAZ-1, AET-1) and the sequential cutting of these wells in the study area. The main conclusions of this study are as follows:

+ The breakdown of the Lower Devonian series shows four genetic sequences for each well. In the Gedinnian, the sedimentary processes at low sea level are well developed compared with those at high sea level, which explains the regression in contrast to the Segennian and Emsian.

+ Some wells have thicker sequences than others.

+ Sedimentation is essentially detritic, and the depositional environment has been interpreted as shoreface.

The isobath maps at the top of the Gedinnian and Siégenien and Emsien show a deepening towards the north-west.

The petrophysical parameters of Devonian reservoirs generally show low porosity and low permeability, which explains tight reservoir character. Cross-plot analysis shows that the correlation coefficients tend towards zero, which explains the poor quality of these reservoirs due to the heterogeneity of the environment.

Statistical analysis shows that the MAMI-2 well has high porosity and permeability values in the Lower Devonian compared with the other wells, confirming a good reservoir.

The 1D geostatistical modelling of petrophysical parameters (porosity and permeability) using Gaussian variograms shows a relatively homogeneous structuring of the results.

Statistical and geostatistical analysis; can be the subject of estimation of these parameters (porosity and permeability) and even others; using kriging (simple or indicative).

The potential reservoirs correspond successively : Ordovician sandstone; Devonian sandstone-siltstone; and Muscovite sandstone-carbonate. These successive reservoirs and a mature Silurian bedrock make the Sillon de Benoud a very interesting petroleum play.

Bibliography

Anadarko. 1993 *Exploration assessment of the Benoud through.* Document SH.

Bolle A. 2000. *Study and consideration of spatial variability.* Revue Française de Géotechnique. 93, 55-66 pp.

Boudjema A.1987.*Evolution structural du bassin pétrolier "triasique" du Sahara nord oriental (Algérie).* PhD thesis, univ. Paris-Sud (Orsay).

Chauvet P. 1999. *Aide-mémoire de géostatistique linéaire application du domaine minier.* Les presses de l'école des mines de Paris, 140p.

Chiasson P., Lafleur J., Soulié M. and Law K. T. 1995. *Characterizing spatial variability of*
a clay by geostatistics. Canadian Geotechnical Journal, 32 (1), 1-10 pp.

Habani H.et Medraoui M.1994.*Analyse structurale d'un tronçon de l'accident Sud atlasique -El*
Kohol (Brezina) - Western Saharan Atlas. Geology engineering *thesis*, USTHB.

Haddoum H.1992.*Etudes structurale des terrains du mole d'In Ouzzal* (Hoggar occidental, Algérie). PhD thesis, USTHB.

Khalfaoui S. 2008. *Evaluation géostatistique des paramètres géotechniques des sols du NordEst algérien.* PhD thesis, Université Badji Mokhtar Annaba - Algeria (2008)

Kazi Tani N.1986.*Evolution géodynamique de la bordure nord-africaine. The North Algerian intraplate domain. Approche méga-séquentielle.* Thèse document, univ.Pau.

Kazi Tani et Durée 1993.*La dynamique du secteur de Bechar depuis le Précambrien.*118 congr.nat.soc,hist. scient. Pau 1993,4 coll. Géologie africaine, pp 137-147.

LF Brown Jr,Fisher WL.1977 M26:Seismic Stratigraphy-Applications to Hydrocarbon Exploration .213-248.

Marache A., Riss J. Gentier S. and Chilès J.-P. (2002). *Characterisation and reconstruction of*
a rock fracture surface by geostatistics. International Journal for Numerical and analytical Methods in Geomecanics; Vol. 26 (9), 873-896 pp.

Mathèron G. 1962a. *Traité de géostatistique appliquée.* Mém. B.R.G.M.,14, Tome I. 333p.

Mathèron G. 1962b. *Traité de géostatistique appliquée.*Mém. B.R.G.M.,14, Tome II. 306p.

SONATRACH internal reports. We quote:

Petroleum potential of the El Ouabed perimeter (Blocks :103 and 313)

Potential of the Hassi R'mel West Benoud hydrocarbon system.

Software used

Petrel version 2017.1

Wariovine version 8

Excel 2016

Summary

The Sillon de Benoud is a transition zone between two major areas, the Saharan Platform in the south and the Saharan Allas in the north.

This furrow is separated from the Atlasic zones to the north by the South Atlasic Accident. This played a major role in structuring the trench into high and low areas.

The Lower Devonian in the study area is essentially detritic, consisting of clayey-gritty deposits.

This study focuses on the petrophysical characterisation of Lower Devonian reservoirs using petrophysical and statistical analysis techniques, and geostatistical modelling. The cores were analysed for the following petrophysical parameters: porosity (0), permeability (K). A total of 428 samples were used in this study. This characterisation shows that the Lower Devonian reservoirs are of poor quality.

Geostatistical modelling shows a relatively homogeneous structure between the two parameters (permeability (K) and porosity (0)).

Key words :

Benoud Trench, North-Western edge of the Saharan Platform, Southern Atlas Accident, Lower Devonian, stratigraphy, petrophysics, geostatistical modelling.

MIX
Papier aus verantwortungsvollen Quellen
Paper from responsible sources
FSC® C105338

Printed by Books on Demand GmbH, Norderstedt / Germany